Commercial Poultry Production on Maryland's Lower Eastern Shore

The Role of African Americans, 1930s to 1990s

Solomon Iyobosa Omo-Osagie II

UNIVERSITY PRESS OF AMERICA,® INC.
Lanham • Boulder • New York • Toronto • Plymouth, UK

University Press of America,® Inc.
4501 Forbes Boulevard
Suite 200
Lanham, Maryland 20706
UPA Acquisitions Department (301) 459-3366

10 Thornbury Road
Plymouth PL6 7PP
United Kingdom

Printed in the United States of America
British Library Cataloging in Publication Information Available

Library of Congress Control Number: 2012933299
ISBN: 978-0-7618-5876-8 (clothbound : alk. paper)
eISBN: 978-0-7618-5877-5

∞™ The paper used in this publication meets the minimum requirements of American National Standard for Information Sciences—Permanence of Paper for Printed Library Materials, ANSI Z39.48-1992

This work is dedicated to my dear wife, Lady Andrea and children Shalom Chidimma, and Keturah Osarieme, for their patience, encouragement, and understanding. I appreciate them more than words can say.

. . . and to the men and women who labor in obscurity and tremendous inconvenience to feed millions of people and enhance their nutritional needs every single day.

Contents

Illustrations vii

Acknowledgments ix

Abbreviations xii

1 Introduction 1

2 Geographical, Historical, and Social Background of Lower Eastern Shore (Somerset, Worcester, and Wicomico) Counties 17

3 Early Phase of Commercialization of Poultry Production, 1930s to 1950s 41

4 Development and Consolidation of Large-Scale Commercial Poultry Production, 1950s to 1990s 67

5 Commercial Poultry Production, Public Health, and the Environment 95

6 Commercial Poultry Production, Working Conditions, and Labor Activism 129

7 Conclusion 155

Bibliography 161

Illustrations

TABLES

2.1 Maryland Census, 1930–1990 28

2.2 Somerset County Population, 1930–1990 28

2.3 Wicomico County Population, 1930–1990 29

2.4 Worcester County Population, 1930–1990 29

2.5 Maryland Population by Race for Selected Years, 1930–1990 32

2.6 Maryland Urban/Rural Population Distribution for Selected Years, 1930–1990 32

2.7 Urban and Rural Population Distributions for the Three Lower Maryland Eastern Shore Counties, 1930–1990 33

3.1 Poultry Farm Operators in Maryland, 1945 53

3.2 Number and Value of Poultry Farms in the Three Lower Eastern Shore Counties, 1945 54

3.3 Illiteracy Rates of the Lower Eastern Shore Counties for People Aged 10 Years and Over (1930) 54

3.4 Attendance and Completion Rates of the Lower Eastern Shore Counties for People Aged 7 to 25 Years Old (1940) 55

3.5 Acreage of Poultry Farms on the Lower Eastern Shore Counties, 1930 56

3.6 Total Rural Farm Population of the Lower Maryland Eastern Shore Counties, 1930 57

3.7 Total Rural Farm Population of the Lower Maryland Eastern Shore Counties, 1940 57

3.8 Poultry Farms and Production on the Lower Maryland Eastern Shore, 1930–1935 58

4.1 Number of Poultry Farms on the Lower Maryland Eastern Shore, 1950–1964 75

4.2 Poultry Sold and Farms Reporting on the Lower Maryland Eastern Shore, 1954–1959 75

4.3 Number of Broilers Sold on the Lower Maryland Eastern Shore, 1954 75

4.4 Rankings of Lower Maryland Eastern Shore Counties with Chickens Sold in the Nation (1978) 77

4.5 Rankings of Lower Maryland Eastern Shore Counties with Chickens Sold in the Nation, 1982 and 1987 78

4.6 Amount of Poultry Sold on the Lower Maryland Eastern Shore, 1992 80

4.7 Delmarva Regional Poultry Statistics, 1999 81

4.8 Racial Breakdown of Census of the Lower Eastern Shore, 1950–1990 83

4.9 Condition of Education on the Lower Maryland Eastern Shore, 1950 85

4.10 Condition of Education on the Lower Maryland Eastern Shore, 1960 85

FIGURE

2.1 Map of the Lower Maryland Eastern Shore 17

Acknowledgments

I must begin by thanking God for His favor. He gave me the mind and diligence to undertake this seminal study. His grace pulled me through the tough times that I encountered in my research for this book.

My deep heart-felt thanks goes to Jeremiah Dibua, Charles Johnson, Jr., Glenn O. Phillips, and Rosalyn Terborg-Penn, all Morgan State University scholars of the highest caliber. They read and critiqued my work and steered me in the direction that made it possible for me to complete this work. Their intellectual and scholarly wisdom and foresight took me farther than I could have imagined. I am indebted to their unwavering support and encouragement. I enormously appreciate the help from Rose Monroe, Carole Quine, Theron Coleman, and Blessing Ogamba, my colleagues at the Baltimore City Community College.

I am also grateful to the following individuals: Patrick Harmon, and Raymond P. Smith, Sr., both retired poultry workers in Pocomoke City and Berlin in Worcester County; the Rev. Jim Lewis, retired Priest and activist; Carole Morison of the Delmarva Poultry Justice Alliance; Elizabeth Stewart, Research Historian at the Maryland Commission on African American History and Culture; University Archives, the Johns Hopkins University, and Ann Hanlon at the Special Collections Department, University of Maryland Libraries. Thanks also to Rebecca Mazur, Research Librarian at the United States Department of Agriculture (USDA), National Agricultural Library, Beltsville, and Sara Lee at the USDA Special Collections.

I am grateful to the archivists and research room assistants at the National Archives, College Park and the Wirtz Labor library at the U.S. Department of Labor, Washington, D.C. They were all very helpful in locating and sorting out relevant data. I cannot forget Cynthia Caldwell and Valda Perry of the Delaware Public Archives for their assistance in locating other useful materials.

My family has been exceptional. I spent many nights alone and in the basement of our home when I should have been with my family doing other “fun” stuff. They put up with my endless absences while I was away on research trip and digging through files and reports of different sorts and conducting field interviews. I thank my wife Andrea for taking care of our home and our children Shalom Chidimma and Keturah Osarieme while I was working on this book. They are phenomenal people and they honored me with their patience. They also provided me with the quiet encouragement that I needed by adjusting to my field research schedule even at the expense of foregoing previously planned family engagements. I love them very much.

Thanks to my late my parents, Chief (Ms.) L. K. Laura Ezekwe and Chief Solomon O. Omo-Osagie, Sr. for their insistence on higher education and commitment and dedication to any task. I am the scholar that I am today because of their influence. I draw my inspiration and motivation from their example. While I miss them, I am confident that they will be pleased with what I have been able to do with what they gave me.

Thanks to everyone who has taught and encouraged me.

Abbreviations

AAA	Agricultural Adjustment Administration
AESES	Agricultural Experiment and Extension Service
BLS	Bureau of Labor Statistics
CERCLA	Comprehensive Environmental Response, Compensation, and Liability Act
CDC	Centers for Disease Control and Prevention
COA	Census of Agriculture
DELMARVA	Delaware, Maryland, and Virginia
DOL	Department of Labor
DPI	Delmarva Poultry Industry, Inc.
DPJA	Delmarva Poultry Justice Alliance
EOS	Environmental Odor Sources
EPA	Environmental Protection Agency
EPCRA	Emergency Planning and Community Right to Know Act
FCA	Farm Credit Administration
FDA	Food and Drug Administration
FFA	Future Farmers of America
FHA	Farmers Home Administration
FSA	Farm Security Administration
FSA	Farm Services Agency
FSA	Food Security Act
FSIS	Food Safety and Inspection Service
GAO	Government Accountability Office
HHS	Health and Human Services
MASS	Maryland Agricultural Statistical Service
MOSH	Maryland Occupational Safety and Health
NAL	National Agricultural Library
NASS	National Agricultural Statistics Service

NFA	New Farmers of America
OSHA	Occupational Safety and Health Administration
PPA	Pollution Prevention Act
PPIA	Poultry Products Inspection Act
RWDSU	Retail, Wholesale, and Department Store Union
TRI	Toxic Release Inventory
UFCW	United Food and Commercial Workers
USDA	United States Department of Agriculture
WPA	Works Progress Administration
WQIA	Water Quality Improvement Act

Chapter One

Introduction

> The Poultry Industry, with its multi-billion dollar annual income to the farmers of the United States, is one of the nation's largest sources of agricultural income. It has not always been big. From a farm operation of relative unimportance it has developed to its present status as a leading source of highly nutritious food to feed a hungry world. This has all happened in the short space of a half century . . .[1]

Poultry[2] has had a consistent presence in America since the early settlers arrived in the New World and began breeding chickens using various genetic manipulations.[3] Over time, domesticated chickens took on a more important role beyond laying eggs to being raised for consumption. The poultry business evolved from being a small and family-centered vocation to a multi-billion dollar industry with sophisticated production techniques such as vertical and horizontal integrations to control output.[4] During the early part of the twentieth century, families raised poultry in their backyards for egg production and to meet their daily protein intake. Poultry production has since become a major source of the food and fiber staple in the nation's food chain with the capacity to supply millions of pounds of poultry meat every week to the nation's food tables.

Americans developed a distinct appetite for poultry meat by World War I. When the United States entered the war in 1917 and soldiers were deployed to the European battlefields, poultry demand rose to meet the needs of the soldiers.[5] The high demand led poultry suppliers to increase production of poultry meat, which helped to boost the morale of the soldiers in the field.[6] In 1928, then Commerce secretary and future president of the United States, Herbert Hoover, who served as the director of the Federal Office of Food Administration during World War I, used the slogan "a chicken in every pot" in his campaign for president. While it cannot be proven that the slogan led to his electoral victory, his adoption of the slogan during the campaign underscored Americans' fondness and preference for poultry consumption.[7]

The importance of poultry to the region is reflected in the value of poultry produced as well as the number of jobs in the poultry industry. According to the Delmarva Poultry Industry Incorporated, in 1956, the Delmarva region's value of poultry produced was $156 million.[8] In the 1960s, 16,736 persons, including poultry growers, farm workers, and processing plant workers, were employed in the poultry industry.[9] During the same period, the industry had a payroll of more than $100 million. In the 1970s, the number of jobs increased to 22,883 with a payroll of more than $149 million.[10] During the 1980s, more than 37,000 persons worked in the poultry industry and had a payroll of $286.6 million.[11] In the 1990s era, the Delmarva region's poultry workforce exceeded 40,000 workers with more than half a billion dollars in payroll.[12] During the same era, the Delmarva region was the fifth largest broiler production area in the United States and accounted for 7% of all poultry produced in the United States.[13]

The poultry industry is a major segment of the Delmarva Peninsula economy. This relatively narrow strip of land 200 miles long between the Chesapeake and Delaware Bays consists of the State of Delaware, the 11 Eastern Shore counties of Maryland, and two counties of Virginia.[14] The close proximity of the three states on the peninsula has removed state line restrictions in commercial poultry production on the Eastern Shore. In the early period of commercial poultry on the Lower Eastern Shore, it was typical for chicken to be raised in Georgetown, Delaware but processed for public consumption by companies located in Salisbury, Maryland. The Lower Maryland Shore is part of an axis of poultry production, a region where poultry is a way of life and the local residents depend on its viability.

Although Delaware, Maryland and Virginia are distinct and unique in terms of culture, values, and orientation, commercial poultry figured prominently in their respective economies. The major poultry force in the region, Perdue Farms, although headquartered in Salisbury, Maryland, has facilities in Maryland, Delaware and Virginia. Mountaire Farms of Delmarva, with headquarters in Delaware, have facilities in Maryland. In essence, these companies had to pay business taxes to all three states and in doing so contributed to their economies. Poultry products continue to move across state lines on a regular basis, making it clearer to understand the integrated nature of commercial poultry production.

The Lower Maryland Eastern Shore was the birthplace of large-scale commercial poultry in Maryland in 1920. Poultry data from the National Agricultural Statistical Service showed a steady increase in poultry production and income for the state's economy and poultry farmers through the various stages of large-scale poultry commercialization. African American involvement in the poultry production on the Lower Maryland Eastern Shore was mainly as laborers. Although the written documents to support their involvement are rare, archival photographs from the trade association, Delmarva Poultry Industry (DPI), as well as oral accounts clearly showed African Americans worked on various poultry farms in the early period of commercial poultry production.

While Maryland has been a poultry producing state for decades, there has been little research done that focused explicitly on the Lower Maryland Eastern Shore with regard to the evolution from small, independent, and subsistent poultry production to large-scale commercial poultry production. The industry on the Lower Maryland Eastern Shore relied heavily on labor provided by an underclass, mainly African Americans. Although the involvement of African Americans in the labor supply-side of the poultry industry is generally known and acknowledged, few written documentation exist to show the extent of their involvement. The poultry companies, which were primarily privately-owned, were reluctant to release their records or information about their workers to researchers and the public. However, several unheard voices made up of former and current workers as well as members of the black community of the Lower Shore lent their voices through oral accounts of their involvement in the poultry industry.

The development of this important industry on the Lower Shore has not received much attention from historians and other scholars. In addition, the scope of African American involvement in the poultry industry has not received much attention. The absence of an earlier historical study of this topic in many ways indicates the presence of a history that deserves documentation. In another sense, the absence arose in part from the lack of interest in agricultural history, which still carries the stigma of slavery.[15] With specific reference to African American students, there has not been sustained interest in the numerous agricultural-centered programs at higher institutions of learning.[16] However, there were exceptions in the early 20th century. In 1928, the Future Farmers of America (FFA), a predominantly white organization, was founded to promote agricultural education throughout the nation. The Maryland branch of the FFA was founded in 1929. But the FFA was founded during the racial tension of the 1920s, and African Americans were denied full participation in the organization by their white counterparts. Consequently, a new organization called the New Farmers of America (NFA) was founded in 1928 by a group of African American farmers to promote vocational agriculture. The NFA flourished and had a membership of over 58,000 before it merged with the mainly white-dominated FFA in 1965.[17]

Nevertheless, the aims and objectives of this work are to show how commercial poultry production evolved on the Lower Maryland Eastern Shore and indicate the extent of African American economic participation. In the process, the importance of commercial poultry production to Maryland's economy will be examined. This will be done by using Agricultural statistics from the Maryland and National Agricultural Statistical Services. This work will also examine the environment, health, and labor practices as they intertwined with commercial poultry production on the Lower Shore.

Large-scale commercial poultry production on the Lower Shore was a major contributor to the local, regional, state, and the nation's economy. But the involvement of African Americans in this industry has largely been understudied. Such an understudy has led to the lack of knowledge of the role of black workers in this important and strategic industry in terms of food production. Such an understudy also belied the role black workers played in enriching whites. For example, in the 1990s, Frank Perdue was named by *Forbes* magazine as one of the richest Americans and the richest person in Maryland.

This work is approached from both thematic and periodization perspectives. Nevertheless, a geographical and historical background of the Lower Shore is necessary and appears in the second chapter. It outlines the landscape of the Delmarva region and examines tobacco and other crops that were prevalent in the region. A noted argument made in this chapter is that of the birth of the commercial broiler industry in the small town of Ocean View, Delaware. In chapter three, an examination of the early phase of commercialization of poultry production from the 1930s to the 1940s follow. The chapter examines the "infancy" period of the industry as one of mixed realities and risk taking. This period ushered in the prospects of commercial poultry industry to become the path to regional economic viability. During this period, poultry brokers from New York and elsewhere established themselves as reliable customers.

In the fourth chapter, the development and consolidation of large-scale commercial poultry production is discussed. It was in the 1950s that large-scale commercial poultry production developed in Maryland with Perdue Farms leading the way. There were also consolidations during the period of the 1950s through the 1990s when small to medium scale poultry companies were purchased by larger companies such as Perdue, Tyson, and Mountaire. After the consolidation of these companies, they were vertically and horizontally integrated, and this enabled the bigger companies to control most aspects of poultry production in the Lower Shore.

The chapter discusses how Delmarva poultry made its mark on the global stage during World War II and the period afterwards. During World War II, the U.S. Armed Forces looked to Delmarva poultry to aid it in ensuring that American soldiers fighting in Europe had their meat and protein ration in the battlefield. Poultry meat's availability served as a morale booster for the soldiers. The fourth chapter also focuses on science and technology. The emphasis on science was mainly on how this industry utilized scientific research to expand and sustain itself for the future. Technologically, the industry began to move in the direction of vertical integration and mechanization of poultry production to ensure high quality and consistency of their product. These processes were intended to improve the ability of the Delmarva region to supply poultry to the consumers with little interruptions.

Chapter five examines the twin issues of the environment and public health and their implications on commercial poultry production. Once large-scale commercial poultry production became dominant, the need to meet the consumer demand increased and intensified the mechanization of poultry production. The demand also meant that large scale companies modified their poultry feed to remain competitive with implications for environmental sustainability.

At issue was the environmental impact that millions of pounds of poultry manure that was generated by large-scale commercial poultry production had on the region. Commercial poultry processing plants used millions of gallons of water every day, and this created a major concern for the preservation of the Chesapeake Bay. On consumer health, scientists and researchers from the Johns Hopkins University, School of Public Health, the University of Maryland, Eastern Shore, and the United States Department of Agriculture (USDA), among others, have conducted ongoing studies and insightful experiments that showed the prevalence of food-borne bacteria such as *salmonella*, *e-coli*, and *campylobacter* in commercial poultry production in Maryland. According to the U.S. Centers for Disease Control and Prevention (CDC), "an estimated 1.4 million cases of *salmonellosis* food poisoning occur [annually] in the United States."[18] Food poisoning became a concern as large-scale commercial poultry production developed.

Chapter six addresses commercial poultry production, working conditions, labor activism, and its labor-intensiveness. From the period of large-scale production, USDA statistics showed that the value and volume of poultry production on the Lower Shore steadily increased over the years. Additional statistics from the U.S. Department of Labor showed that workers' wages did not increase in proportion to the level of profits of the large-scale companies around the country. Likewise, the working and safety conditions barely improved. Surveys conducted by the Occupational Safety and Health Administration (OSHA) and the Department of Labor showed that poultry company workers were more likely to sustain injuries on the production line than other manufacturing and industrial workers.

Safety concerns and working conditions led the Department of Labor and OSHA to institute stricter workplace rules and regulations in the poultry processing plants during the 1990s. For example, companies were required to provide workers with alternative and ergonomically acceptable machine operations to minimize repetitive motion injuries on the production lines as well as slip and fall accidents. These workplace injuries in large-scale commercial poultry production drew the activism of the labor unions such as the United Food Commercial Workers (UFCW) and the Retail, Wholesale, and Department Store Union (RWDSU) began advocating for safer working conditions and higher wages for poultry workers on the Lower Maryland Eastern Shore. The Delmarva Justice Poultry Alliance (DPJA), a pro-environmentalist conglomeration of seventeen organizations and whose main focus was workers' rights to fair wages, safer working conditions, and environmental preservation, was founded in 1999. Chapter seven concludes with a summary and an analysis of the involvement of African Americans in commercial poultry production on the Lower Maryland Eastern Shore.

The existing literature has typically offered a general overview and background of the history of the Delmarva region. In the secondary literature on the Eastern Shore, there was little to no emphasis on the role of African Americans in the poultry industry. The available literature has focused mainly on its overall importance to the region's economy. Yet large-scale commercial poultry was a major source of employment for African Americans in this region.

Much has been written about the Eastern Shore region's economic viability and prospects since the construction of the Chesapeake Bay Bridge in 1952. The construction of the bridge was crucial for the region's economy as it linked it with the rest of the state and enabled businesses, including Perdue Farms, to transport their poultry to other parts of the state.[19] Historians George H. Corddry, Vera Foster Rollo, and John Wennersten among others have written generally about the histories of Wicomico County, as well as the black experience in Maryland, and the Maryland Eastern Shore. These works only made general references to poultry as an important segment of the local economy and do not focus on the involvement of African Americans in the industry.[20]

Commercial poultry production on the Maryland counties of the Delmarva Peninsula is for the most part a story about Perdue Farms. There are only two books that were written on poultry pioneer Arthur Perdue, *A Solid Foundation: The Life and Times of Arthur W. Perdue*, by Frank Gordy, published in 1976 and *Frank Perdue: Fifty Years of Building on a Solid Foundation* by Mitzi Perdue, published in 1989. Gordy's book was specifically authorized by Arthur Perdue himself and contains a wealth of information on the Perdue family dating back to the 1600s. According to Gordy, Henri Perdue, a French Huguenot, arrived in Maryland from France with his family to take advantage of Maryland's religious tolerance. The Act of Religious Tolerance of 1649 gave freedom of worship to English Protestants and Catholics who wanted a place of religious freedom. The Act gave succor to those who were fleeing from religious persecution. They settled in the Maryland colony. Upon their arrival in Maryland, Henri settled with his family in Worcester and Wicomico Counties, which became home to many generations of Perdue.[21] Gordy's account of Arthur Perdue painted him in most personal terms. The personal depiction sheds light into the culture and values of Perdue Farms. Gordy depicted him as a tenacious man, a characteristic that appeared to have contributed to his notoriety as the inventor of large-scale commercial poultry production on the Lower Maryland Eastern Shore.

The book by Mitzi Perdue, Frank Perdue's widow, also contains pertinent information. It was written in celebration of the 50th anniversary of the Perdue Farms' founding. A major distinction between Frank Gordy and Mitzi Perdue's books is that Mitzi Perdue's work focused on a timeline of important landmarks in the company's history beginning in 1920 when Arthur Perdue resigned his job with the railway company to 1989 when Perdue Farms was the dominant poultry industry in the region. Like Gordy's publication, Mitzi Perdue's volume has useful primary documents because the sources came from the Perdue family archives and personal collections.

At the time of its publication by the Arthur W. Perdue Foundation, Mitzi was married to Frank Perdue. This family connection raised legitimate questions about the work's objectivity and whether the work can withstand scholarly scrutiny. Nevertheless, researchers and scholars can benefit from the important information on Frank Perdue contained in the book. It is imperative, however, to use the book from the perspective of its lack of historical and professional review partly because of Mitzi Perdue's background as an untrained historian.

A work by William H. Williams entitled *Delmarva's Chicken Industry: 75 Years of Progress*, considered the most significant work to date, focused on the poultry industry on the Delmarva region. It was published in 1998 by the Delmarva Poultry Industry, Inc (DPI). This work was also written from a sympathetic perspective and showcased the "progress" of the poultry industry. It contains useful analyses of the industry and the economic vibrancy that it has brought to the region, particularly its transformation of an otherwise economically depressed region. But the book came short of fully and adequately treating environmental and public health issues.

Williams' work is mainly a broad sweep of the poultry industry on the entire Delmarva Peninsula with emphasis on the Delaware part of the Delmarva, which he claimed was the birthplace of commercial poultry production in the region. A notable observation of this work is that it was sponsored and published by the Delmarva Poultry Industry Incorporated, the same organization that serves as the trade group that represents the interests of the poultry industry. Clearly, the narrative was written from an advocacy standpoint. In addition, the book did not provide much in-depth discussion about the vast environmental and public health issues, which particularly intersected with commercial poultry production. However, the focus of this current work provides a full and in-depth examination of the origins, evolution from small scale to large scale commercialization of poultry, public health, environmental protection, working conditions, and labor issues involved in commercial poultry production on the Maryland side of the Delmarva.

Noticeably, there had been a general scholarly interest in this topic. Several theses, dissertations, and scholarly articles that were examined for this work, either partially or tangentially, addressed the central focus of this current book.[22] In most of these works, the same issues of race, gender, and class figured prominently in the poultry, beef, and hog commercial production mainly because of the preponderance of the underclass as the agency for meat production in the United States.[23] For example, Roger Horowitz's work examined meatpackers. Although his work focused on the Chicago area meatpacking houses, the United Packinghouse Workers of America (UPWA's) activities mirrored those of the United Food and Commercial Workers (UFCW) to which many of the commercial poultry workers on the Lower Maryland Eastern Shore belonged. Another work by Eric Brian Halpern, "'Black and White Unite and Fight': Race and Labor in Meatpacking, 1904–1948," also focused on the Chicago meatpacking industry. Halpern's work revealed that race was an important factor in unionization. This study also revealed a trend whereby white meatpacking workers were the first to leave the industry upon gaining an opportunity to move into other industrial jobs.[24]

Conversely, blacks moved into the meatpacking positions left by whites' upward mobility into higher paying and more "respectful" jobs. But the replacement occurred after blacks and whites had formed an interracial coalition that made it possible for the UPWA to mount an aggressive campaign that led to better workplace conditions for meatpackers. There was also a class commonality between poor blacks and poor whites. The labor in the meatpacking houses was provided mainly by the underclass. Likewise, on the Lower Maryland Eastern Shore, there were whites who worked on the labor supply-side of commercialized poultry production but left after other industrial jobs became available. Local blacks moved into these vacant positions, thereby creating both a black working class and an underclass.

Horowitz and Halpern's works provided a contextual framework that infused race, class, and gender into commercial poultry production on the Lower Maryland Eastern Shore. Their works highlighted race and class formations and underscored the significance of subaltern theory in working class historiography, as blacks were instrumental in identity construction and formation regarding their work in the meatpacking and commercial poultry industries.[25] What the subaltern theory indicates in striking terms is that the voices of the underclass are central to documenting the history of commercial poultry production on the Lower Maryland Eastern Shore. These voices help answer crucial questions about the working conditions in the processing plants and the low wages of the workers in spite of the documentary evidence that showed the industry's profitability.

On their part, black women have been involved in poultry processing and meatpacking industry since World War II when the men, many of whom were the breadwinners of their families, left to fight in the battlefields of Europe. The masculinity of the labor workforce threatened to overshadow the contributions that black women in particular had made to the industry. Bruce Fehn argued that such masculinity was symptomatic of the nature of African American involvement in the commercial meat production as black women constituted a major part of meat production and processing plant line work.[26]

Another important work by Debra Fink, *Cutting into the Meatpacking Line: Workers and Change in the Rural Midwest*, showed a connection or direct relationship between a rural working underclass and commercialized food production.[27] The setting of Fink's book, a rural Iowa community, bears a resemblance to the Lower Maryland Eastern Shore: rural, small population and a rich agricultural landscape with a less educated workforce where a majority of the residents are farmers or are engaged in some form of agriculture.

Several notable studies examined the intersection between commercial poultry production, the environment, and public health. These intersections indicated the existence of a struggle to balance profit considerations with public health needs and environmental sustainability. In one of these studies, D.J. Kraft, Carolyn Olechowski-Gerhardt, J. Berkowitz, and M.S. Finstein found that *Salmonella* was present in 26 out of 91 poultry houses in sampled New Jersey commercial poultry farms and that *Salmonella* was present in streams near poultry farms. The presence of *Salmonella* in the water had been a concern to the surrounding communities because many of the local residents made their living as watermen and women.[28] Another study showed that although *Salmonella* existed in the streams, an elevated temperature technique can be used to isolate it from streams.[29] The connection of these studies to the Lower Eastern Shore lies in the numerous streams and creeks in and around many commercial poultry farms and processing plants. The presence of *Salmonella* would have the same impact on all large-scale commercial poultry operations.

The Chesapeake Bay was held up by many researchers as an example of the harmful effects of commercial poultry production on the environment. Studies by scientists, public health specialists, and environmentalists who have focused on the Bay have consistently argued that the run off from the commercial poultry processing plants have impacted communities in and around the Bay and have damaged maritime life and made many waterborne bacteria resistant to antibiotics and water treatment. Minorities have borne the ill effects of environmental polluting water run-offs from the poultry processing plants located in the heart of black or Latino communities.[30] For example, the Perdue poultry processing plant is located in the heart of the black community of Salisbury while the corporate headquarters is located several miles away. In each instance, environmentalists have faulted the industry's treatment of the Bay and its impact on the surrounding communities.

The significance of large scale commercial poultry production was not lost on public officials, especially those who represented the poultry producing regions such as the Lower Maryland Eastern Shore. Government Accountability Office (GAO) reports and floor speeches in the United States House of Representatives, the United States Senate, the General Assembly of Maryland, and comments offered during floor debates in the legislatures were indicative of the prominence of poultry. Much of the comments related to workplace conditions, safety regulations, wages, public health and environmental concerns.[31]

For example, the former U.S. Representative Roy Dyson, who represented the Lower Maryland Eastern Shore in the House during the 1980s, articulated the region's poultry interests on the House floor as well as within the agricultural committee and sub-committees. On the Senate side of the U.S. Congress, Maryland Senator, Barbara Mikulski also made speeches on the Senate floor regarding poultry, especially the Poultry Products Inspection Act [Public Law 85-172] enacted in 1957 to regulate the production and dissemination of poultry products for public consumption and the efforts to amend sections of the law by Congress. Senator Mikulski's supportive remarks about the economic impact of Delmarva poultry production boosted the morale of the poultry industry, which had received much criticism from animal rights groups. The passing of Frank Perdue and the economic contributions that he made not only to the Lower Maryland Eastern Shore, but also to the U.S. economy through job creation and philanthropic works were particularly noted on the U.S. Senate floor.[32]

There were also poultry legislative interests in the General Assembly of Maryland. For example, in 1937, the Maryland House of Delegates established and empowered the Maryland Board of Agriculture to supervise the sale and transportation of poultry in Maryland. The Board was also empowered to issue licenses to those seeking to operate poultry businesses and impose fines on those who broke the laws on poultry production in Maryland. One piece of legislation on poultry called for a $100.00 fine and imprisonment of 30 days for anyone who purchased poultry from an unlicensed broker or company.[33] In other words, the Maryland State government identified commercial poultry as a potential economic resource much earlier on that it enacted legislation to regulate the enterprise.

Although by various oral accounts, blacks played a significant role in the commercial poultry processing plants labor supply, local media, with the exception of the [Salisbury] *Daily Times*, gave scant coverage to blacks' involvement in the commercial poultry industry.[34] Instead, the local newspapers covered poultry more generally as local news of interest during the years of poultry industrial expansion in the 1950s. However, in later years, during the 1980s and 1990s, and with the influx of low-paid foreign and immigrant workers, national newspapers such as the *Baltimore Sun*, the *Washington Post,* and the *New York Times* devoted more coverage to the involvement of black and Latino poultry workers in the industry.[35]

A review of the 1930 and subsequent censuses showed that the USDA collected data on agriculture. Much of the information in these censuses was confidential to avoid disclosing individual poultry company's data. The census agricultural information, therefore, did not assemble data specifically on black commercial poultry workers or farmers. Specific and more detailed data on black poultry workers began with the 1970 census when the data collection methods were modified to capture racial data. The absence of racial information in the census data prior to the 1970s necessitated the need for alternative sources in examining the involvement of African Americans in large-scale commercial poultry production on the Lower Maryland Eastern Shore.

Consequently, the use of extended oral interviews of former poultry workers and local and knowledgeable individuals helped place their involvement and contributions in the proper context. The eyewitness accounts further the understanding and appreciation of the workers' impact on the industry as well as the industry's impact on the workers and their families. These accounts also revealed the relevance of the poultry industry to the black community on the Lower Maryland Shore since the 1930s until the face of the poultry working class and underclass changed with the influx of migrant Latinos from the 1980s who then took over some of the work that was previously performed by African Americans.

This work employs the subaltern theoretical framework in examining the involvement of African Americans in commercial poultry production on the Lower Maryland Eastern Shore. It attempts to place their involvement in the context of the evolution and transformation of commercial poultry production in the Lower Maryland Eastern Shore. The subaltern framework allows blacks as the original actors to tell their own stories first hand. Hearing directly from the working class or the "rank and file" workers and their perspectives, juxtaposed official accounts with the actors' accounts, which varied due to certain factors and conditions. Some of the data that was collected from poultry production by public servants differed sharply from the accounts of the primary actors.[36]

The subaltern framework brings out the voices of the frontline workers who made the commercial poultry industry into a profitable venture. Their compelling accounts tell of family and individual sacrifices in the face of racial tension, persistence, hard work, and an indomitable spirit to provide for their families against all odds. The subaltern framework in the most striking terms placed African Americans in the center of the growth and expansion of the industry on the Lower Shore. Historically, blacks' presence on the Shore dated over two hundred years. The preceding generations found creative ways to survive racial and social difficulties. In the modern period, however, blacks' involvement in the industry shifted from chicken catching to poultry plant processing. Throughout, their involvement was confined to the labor-supply side and not the ownership ranks because of the lack of access to the credit and capital that was required to undertake such a major business risk. Beginning in the 1980s through the 1990s, an influx of Hispanics and Caribbean workers led to a gradual change of the face of Maryland commercial poultry production.

In summary, this work traced the beginnings and development of commercial poultry production on the Lower Maryland Eastern Shore up to the 1990s and the involvement of African Americans in the industry. African Americans were mainly involved in poultry production on the labor supply side, which was crucial to the expansion of the industry. After it became commercialized in the 1930s and showed great promise in the immediate post-World War II years, poultry production expanded and became the dominant economic activity on the Lower Maryland Eastern Shore from the 1950s. The industry expanded through innovative ways such as vertical integration, acquisitions, mergers, and consolidations.

The industry intersected with public health and the environment. The public health implications arose from the introduction of medications in chicken feed, which negatively impacted consumers and caused poultry-borne infections and diseases such as *Campylobacter*, *Listeria*, *Salmonella*, *E. coli*, *Pseudomonas aeruginosa*. In the environmental sphere, commercial poultry production led to water contamination, air pollution, and land degradation. These intersections were problematic for the industry as it attempted to balance a needed and important industry that was crucial to the economic lifewire of the region on the one hand, and on the other, protect public health and ensure a sustainable environment.

Despite large profits accumulated by the industry, issues such as fair wages and working conditions dominated the interactions between the poultry industry and the workers. The result was a labor activism led by the Delmarva Poultry Justice Alliance (DPJA) and the United Food and Commercial Workers Union (UFCW) that forced poultry companies to confront and deal with the workers' issues. Their activism ultimately helped to bring about changes in wages and working conditions. In the final analysis, an industry that began on a very small scale metamorphosed into a large-scale behemoth that changed the eating habits and the lives of a region, state, and nation with significant impact on the environment and public health.

NOTES

1. John L. Skinner, (ed.), [Statement of Belief] quoted in *American Poultry History, 1823–1973* (Madison, Wisconsin: American Printing and Publishing Inc., 1974), 608.
2. In this work, I will be using poultry and broilers interchangeably. Wherever they are used in the book, I am referring to chickens.
3. Glenn E. Bugos, "Intellectual Property Protection in the American Chicken-Breeding Industry," *The Business History Review*, Volume 66, No. 1 (Spring, 1992), 127–168.
4. Ewell Paul Roy, "Effective Competition and Changing Patterns in Marketing Broiler Chickens," *Journal of Farm Economics* Volume 48, Issue 3 (August, 1966), 188–201.
5. Ben Kerner, *Your Chicken Has Been To War* [motion picture documentary]. Dover: Delaware State Archives, Division of Historical and Cultural Affairs, Hall of Records, 1943.
6. Gordon Sawyer, *Agribusiness Poultry Industry* (Jericho, New York: Exposition Press, Inc., 1971), 76.
7. Richard T. Rogers, "Broilers—Differentiating a Commodity," Unpublished paper (n. d.), 5–12.
8. "Delmarva's Chicken Business is Important TO YOU!" Delmarva Poultry Industry, Inc., 1957.
9. "Facts About Delmarva's Broiler Industry," Delmarva Poultry Industry, Inc., (DPI) 1961 and 1965 (DPI Archives). The DPI also published industry facts and

figures every five years during the periods covered in this study: 1970; 1975; 1980; 1985; 1990, and 1995.

10. Ibid.

11. Ibid.

12. Ibid.

13. Ibid., and the *National Agricultural Statistical Service*, United States Department of Agriculture (USDA).

14. "The Delmarva Peninsula—Birthplace of Commercial Broiler Industry," Delmarva Poultry Industry, Inc. [pamphlet—n. d.].

15. Eric St. John, "Retiling the Fields," *Black Issues in Higher Education*, (8 June 2000).

16. Joan Morgan, "African Americans and Agriculture," *Black Issues in Higher Education* Volume 17, Issue 8 (June, 2000).

17. The photographic images of NFA activities, African Americans, newspaper clippings, letters, and guides are housed at the Virginia Tech Special Collections Department. See also, Dexter B. Wakefield and B. Allen Talbert, along with references, "Exploring the Past of the New Farmers of America (NFA): The Merger with the FFA," *Proceedings of the 27th Annual National Agricultural Education Research Conference*, [n.p., n.d.], 420–433.

18. "Poultry Pathogens: Models for Predictive Microbiology," *Agricultural Research* (June, 2005).

19. See the following works, Eric Van de Verg (ed.,) and John H. Cumberland (ed.,), *Proceedings: Second and Third Annual Conferences on the Economics of Chesapeake Bay Management*. Annapolis, Maryland, 28–29 May [1986], and 27–29 May 1987.

20. See George H. Corddry, *Wicomico County History*. (Salisbury, MD: Peninsula Press, 1981): Vera Foster Rollo, *The Black Experience in Maryland*. (Lanham, MD: Maryland Historical Press, 1980), and John Wennersten, *Maryland's Eastern Shore: A Journey in Time and Place*. (Centerville, MD: Tidewater Publishers, 1992).

21. Frank Gordy, *A Solid Foundation* . . . 3–9.

22. See the following examples, Papers of Morley A. Jull, Series 1: Poultry Husbandry, Box 1, folder 6, [Working Papers, n.d.]; Homer W. Walker and John C. Ayres, "Incidence and Kinds of Microorganisms Associated with Commercially Dressed Poultry," *Applied Microbiology*, Volume 4, No. 6 (November 1956), 345–349, P. Morris, et al., "Respiratory Symptoms and Pulmonary Function in Chicken Catchers in Poultry Confinement Units," *American Journal of Industrial Medicine*, Volume 19 (1991), 195–204. See the following theses and dissertations at the University of Maryland, College Park, C. W. Pierce, "An Economic Study of 99 Maryland Poultry Farms,"(M.S., thesis, 1933); Paul R. Poffenberger, "An Economic Study of the Broiler Industry in Maryland," (M.S., thesis, 1937); William J. Lodman, "Farm Credit on the Lower Eastern Shore of Maryland," (M.S., thesis, 1941); Thomas Joseph Davies, "The Broiler Industry in Maryland," (M.S., thesis, 1942); John Stanley Stiles, "Comparative Costs of Cutting and Packaging Poultry," (M.S., thesis, 1958); George Edward Reid, Jr., " An Analysis of the Market Organization and Structure of the Maryland Table Egg Industry," (M.S thesis, 1966); Olivier Lafourcade, "Analysis

of Optimal Marketing Strategies for the Poultry Industry in Delmarva," (Ph.D. dissertation, 1971); Donald L. Van Dyne, "Economic Feasibility of Heating Maryland Broiler Houses with Solar Energy," (Ph.D. dissertation, 1976), and Lewell F. Gunter, "The Optimal Replacement of Breeder Flocks in an Integrated Broiler Farm," (Ph.D. dissertation, 1979). See also other works by, Bill R. Miller, Ronaldo A. Arraes and Gene M. Pesti, "Formulation of Broiler Finishing Rations By Quadratic Programming," *Southern Journal of Agricultural Economics* (July, 1986), 141–150, Charles R. Knoeber and Walter N. Thurman, "Testing the Theory of Tournaments: An Empirical Analysis of Broiler Production," *Journal of Labor Economics*, Volume 12, No. 2 (1994), 155–179, and Alan Barkema and Mark Drabenstott, "The Many Paths of Vertical Coordination: Structural Implications for the US Food System," *Agribusiness*, Volume 11, No. 5 (1995), 483–492.

23. See for example, Roger Horowitz's dissertation, "The Path Not Taken: A Social History of Industrial Unionism in Meatpacking, 1930–1960," (Ph.D. Dissertation, University of Wisconsin-Madison, 1990), which showed a strong presence of the underclass in the meatpacking industry as was seen in the United Packinghouse Workers of America (UPWA).

24. Eric Brian Halpern, "'Black and White Unite and Fight': Race and Labor in Meatpacking, 1904–1948," (Ph.D. Dissertation, University of Pennsylvania, 1989).

25. See Roger Horowitz and Rick Halpern, "Work, race, and identity: self-representation in the narratives of black packinghouse workers" *Oral History Review*, 26:1(1999), 23–43.

26. See Bruce Fehn, "African American women and the struggle for equality in the meatpacking industry, 1940–1960," *Journal of Women's History*, 10:1 (1998), 45–69.

27. Deborah Fink, *Cutting into the Meatpacking Line: Workers and Change in the Rural Midwest* (Chapel Hill: University of North Carolina, 1998). See also the following works, Eric Arnesen, "Up From Exclusion: Black and White Workers, Race, and State of Labor History," *Review in American History*, Volume 26, No. 1 (March 1998), 146–174, Rick Halpern and Roger Horowitz, *Meatpackers: An Oral History of Black Packinghouse Workers and their Struggle for Racial and Economic Equality* (New York: Twayne Publishers, 1996); Rick Halpern, *Down on the Killing Floor: Black and White Workers in Chicago's Packinghouses, 1904–1954* (Urbana: University of Illinois Press, 1997); Shelton Stromquist and Marvin Bergman (eds.), *Unionizing the Jungles: Labor and Community in the Twentieth-Century Meatpacking Industry* (Iowa City: University of Iowa Press, 1997).

28. D.J. Kraft, Carolyn Olechowski-Gerhardt, J. Berkowitz, and M.S. Finstein, "Salmonella in Wastes Produced at Commercial Poultry Farms," *Applied Microbiology*, Volume 18, No. 5 (November, 1966), 703–707.

29. Donald F. Spino, "Elevated-Temperature Technique for the Isolation of Salmonella From Streams," *Applied Microbiology*, Volume 14, No. 4 (July, 1966), 591–596.

30. See the following works, Robin Marks, "Cesspools of Shame: How Factory Lagoons and Sprayfields Threaten Environmental and Public Health," (Washington, D.C.: Natural Resources Defense Council and the Clean Water Network, 2001), 24; Intensive Poultry Production: Fouling the Environment," *United Poultry Concerns*

(n.a and n.d.); Doug Parker, "Alternative Uses of Poultry Litter," *Economic Viewpoints*, Volume 3, No. 1 (Summer, 1998); David Morris and Jessica Nelson, "Looking Before We Leap: A Perspective on Public Subsidies for Burning Poultry Manure," *Institute of Self-Reliance* (n.d.); William C. Baker and John D. Groopman, "Health of the Bay—Health of People Colloquium [Introduction]," *Environmental Research Section* A 82, (2000), 95–96; Leo Horrigan, Robert S. Lawrence, and Polly Walker, "How Sustainable Agriculture can Address the Environmental and Human Health Harms of Industrial Agriculture," *Environmental Health Perspectives*, Volume 110, Number 5 (May, 2002), 445–456; David F. Boesch, "Measuring the Health of the Chesapeake Bay: Toward Integration and Prediction," *Environmental Research Section* A 82, (200), 134–142; Peter L. deFur and Lisa Foersom, "Toxic Chemicals: Can What We Don't Know Harm Us,?" *Environmental Health Section* A 82, (2000), 113–133, and Timothy Maher, "Environmental Oppression: Who Is Targeted for Toxic Exposure?" *Journal of Black Studies,* Volume 28, No. 3 (January, 1998), 357–367.

31. United States Government and Accountability Office. Report to the Ranking Minority Member, Committee on Health, Education, Labor, and Pensions. "Workplace Safety and Health: Safety in the Meat and Poultry Industry, while Improving, Could Be Further Strengthened." United States Senate [January 2005].

32. *Congressional Record*—Senate [S3195], 5 April 2005.

33. *The Annotated Code of the Public General Laws of Maryland*, 1930; Volume 379, Chapter 347, Sections 147–152 [Maryland State Archives].

34. The local newspapers included: *Somerset News, Worcester Democrat, Crisfield Times, Marylander and Herald, Somerset Herald, Wicomico News, Worcester Democrat and Ledger Enterprise, Worcester County Messenger, Maryland Times Press, and the Eastern Shore Times,* among others.

35. See for instance articles by Lena H. Sun and Gabriel Escobar, "On Chicken's Front Line," *Washington Post*, 28 November 1999, Kate Shatzkin, "Perdue Sued by Chicken Catchers," *The Baltimore Sun*, 19 September 1998, http://www.poultry.org/labor, and David Barboza, "Meatpackers' Profits Hinge on Pool of Immigrant Labor," the *New York Times*, 21 December 2001.

36. See Eric Arnesen's critical Subaltern work, "Up From Exclusion: Black and White Workers, Race, and the State of Labor History," *Reviews in American History*, Volume 26, Number 1 (March, 1998), 146–174, in which he assailed industrial relations scholarships that focused more on institutions instead of incorporating the voices of the rank and file workers into the labor historiography.

Chapter Two

Geographical, Historical, and Social Background of Lower Eastern Shore (Somerset, Worcester, and Wicomico) Counties

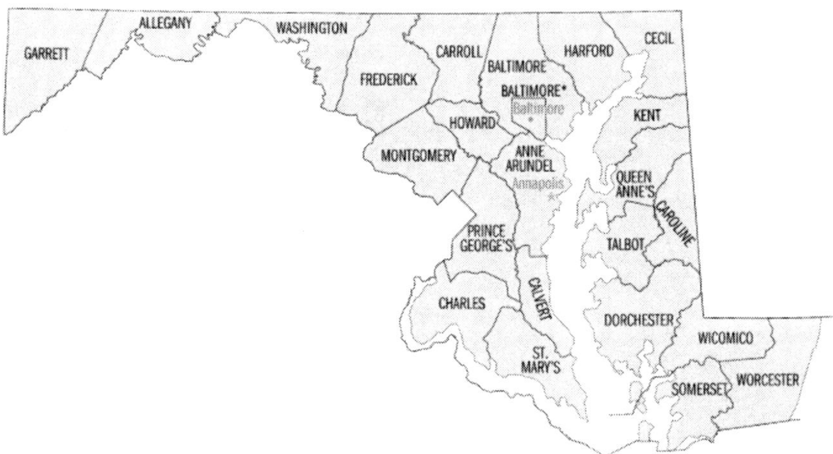

Figure 2.1. Map of the Lower Maryland Eastern Shore. Source: U.S. Census Bureau [Maryland County Selection Map], Available at http://quickfacts.census.gov/qfd/maps/maryland_map.html

GEOGRAPHICAL BACKGROUND

Maryland is one of America's Border States. It is neither a fully northern nor a fully southern state. The phrase "Eastern Shore" was coined for the convenience of the early explorers to describe the area beyond the Chesapeake Bay. James C. Mullikin has argued that "[Eastern Shore] has no official standing, nor has the name Eastern Shore as applied to the trans-Chesapeake counties of Maryland and Virginia.[1] The term not only refers to its geographic position but also to the culture and history of the State as it relates to industries and slavery. However, since the phrase Eastern Shore is conventionally and generally used to describe the area, this study will adopt the terminology.

Geographically, Maryland covers a landscape of 12,192.97 square miles. Of this landscape, 9,843.62 square miles is land based while 623.35 square miles is inland water. The Chesapeake Bay, "the nation's largest bay,"[2] covers 1,726.0 square miles. The state is divided into five provinces. These provinces stretch from the eastern and western regions with varying mountains and altitudes. In addition, the provinces are described as follows:

> The area beyond the Chesapeake Bay and the Atlantic Ocean constitute the Coastal Plain while Catoctin Mountains in Washington County include the Piedmont region. On the western side of the Shore is the Blue Ridge province, which comprises most of Fredrick County. The Valley and Ridge Province, areas including Washington and Allegany County, and Appalachian Plateau Province extend from the base of Catoctin Mountains to western boundary of the State.[3]

The Lower Eastern Shore Counties of Somerset, Wicomico, and Worcester cover a combined total landscape of 1,325.62 square miles (about 10.9 percent) of the entire state square mileage. The three counties also constitute 160.25 square mileage of inland water. Somerset County has 338.41 square miles of land with 35.85 of it being water area. It is a rural and water-oriented county and lies on 46 feet above sea-level. Its major towns include Crisfield and Princess Anne. The county averages 43.4 inches of rain and 11.6 inches of snow per year. The temperature averages 74.4°F in the summer, and 38.4°F in the winter season.

In short, the geographical terrain of Somerset County made it an agricultural county, something that it has remained for over three centuries since its founding. In particular, the total population of African Americans in this county for the years under study gives insight to the presence of African American labor even before commercial poultry production became dominant. Blacks became a major source of labor for tobacco production. Tobacco was important because of its international appeal. Its importance therefore made African Americans vital contributors to the Lower Shore's economy. Tobacco's international demand also helped local planters who traded with European merchants and whose trade routes intersected between Maryland and Virginia. Tobacco remained the dominant crop in Maryland for over a century.[4]

Wicomico County is the largest of the three counties and is located on 379.10 square miles. Salisbury, the seat of the county's government, is the largest city on the Lower Maryland Eastern Shore. The county averages 44.8

inches of rain and 11.4 inches of snow annually. Summer temperature averages 86°F while winter temperature averages 25°F. Its favorable climate and agricultural outlook made the county a preeminent producer of livestock, poultry, and poultry products. In addition, the county is an important producer of fruits and vegetables.[5]

Worcester County is the only county on the Maryland Lower Eastern Shore that borders the Atlantic Ocean.[6] The county has a total area of 584.86 square miles with 474.86 of land and 110 square miles of water. There are three major U.S. highways that run through the county: Routes 13 and 113 connect Wilmington, Delaware, and Norfolk, Virginia, while Route 50 provides access to Baltimore, Annapolis, and Washington, D.C. By the late nineteenth century, the county was a major producer of fruits, vegetables, grains, and lumber.[7] However, by the twentieth century, Ocean City, a major tourist attraction in the region, became a crucial source of revenue for the county as it attracted a steady stream of tourists each year.

Thus, with regard to the geography of the Lower Maryland Eastern Shore, the physical characteristics of the three counties made them ideal for agricultural farming. These characteristics include favorable and non-extreme climate and a fair balance between in the distribution of land and water. Another important characteristic is precipitation. It is a crucial climatic resource and contributes to water supply that is necessary to meet the agricultural needs and focus of the region. Precipitation has particularly been helpful to the region's agricultural viability. The arability of the soil is another important characteristic of the three counties.

HISTORICAL BACKGROUND

African Americans have been a part of Maryland history since the vessels the *Ark* and the *Dove* arrived in Maryland in 1634 with slaves from Africa and the Caribbean.[8] It was a turbulent journey for the slaves who spent over three months on the high seas, having departed London on 22 November 1633 and arrived at Point Comfort on 27 February 1634. One of the vessels, the *Dove,* returned back to its point of origin in London three days after departure due to intemperate weather conditions but reappeared about eight weeks later in Barbados.[9] According to the passengers' manifest, early known Africans in Maryland—Mathias Sousa, John Price, and Francisco—were believed to have arrived with the English settlers aboard the *Ark* and the *Dove*.[10] They were transported to Maryland as servants to the aristocratic class as well as indentured servants to the wealthy emigrants who came on the *Ark* and the *Dove*. Their servant and indentured servants status eventually gave way to enslavement, which solidified and helped to sustain the aristocratic class. The white aristocrats built elaborate mansions and estates with black labor and used black laborers to preserve a luxurious lifestyle. Throughout the Lower Eastern Shore, the white aristocratic class used their position and power to create an economic order that elevated the racial tension between whites and blacks. The racial tensions remained in place long after the aristocratic influence waned. In place of the aristocratic influence, local whites maintained residential segregation, a strict racial code, especially in Somerset County, which helped further racial tension on the Eastern Shore.[11]

The prominence of the Atlantic slave trade in the New World gave impetus to European nations' quest to establish new colonies. The captured Africans were sent to North America and the Caribbean colonial plantations. Slavery persisted in Maryland during the eighteenth century. Many of the slaves who went to the West Indies from Africa were sent to Maryland.[12]

Historically, slaves began arriving in larger numbers on the Eastern Shore in 1685.[13] Before and during their arrival, Maryland condoned and sanctioned slavery, which lasted well into the antebellum period. The permissiveness of slavery gave merchant-planters a free reign over both indentured servants and slaves in the sense that whites were at liberty to determine the nature and volume of labor to be provided. The planters also had disciplinary power, which they used to impose order and control in terms of minimizing work-related interruptions such as slave runaways and idling. The planters had a vested interest in maintaining and preserving the "peculiar institution" because it was essential to their economy and survival.[14]

During the Civil War, slaves were caught in the middle of the debate between Unionists and Confederates. Slaves on the Lower Maryland Eastern Shore were much sought after. The Union government recruited both free blacks and fugitive slaves to fight in the War. In some instances, the Union government paid slave holders for their slaves. For example, Isaiah Fassett, a slave owned by one Sarah A. Bruff, was released in 1863 after the U.S. Army paid $300 to purchase Fassett from Bruff. Maryland did not take kindly to Fassett's purchase by the U.S. Army and was insulted by the action.[15] In particular, slaveholders were defiant of the Union's efforts to recruit black slaves and saw such efforts as attempts to empower slaves. Slavery was a particularly sore subject on the Shore. For two centuries before the Civil War, the white elite had relied on slave labor to create and maintain a high social status. The practice of slavery flourished on the Eastern Shore because of the existence of a plantation economy.[16] The existence of large plantations on the Shore and their benefits to white merchant-planters helps to explain the vociferous opposition to attempts to empower or free more slaves.

Slavery remained entrenched in Maryland until after President Abraham Lincoln's Emancipation Proclamation and Maryland abolished the institution of slavery with the ratification of its own 1864 Constitution. Article 24 of the Declaration of Rights in 1867 reiterated that ". . . slavery shall not be reestablished in this State . . ."[17] The Declaration reinforced the 1864 Constitution in which Maryland mandated freedom for all blacks in the state.[18]

In the same year, the Maryland Legislature met in a contentious session in Annapolis to consider proposals for a constitutional convention. A dispute ensued about selecting delegates. Only those who were loyal to the Union were deemed eligible as convention delegates. This created a problem for Eastern Shore legislators who represented a region with sustained anti-Unionist sentiments.[19]

The legality of slavery in Maryland seemed to have emboldened Unionists. Perhaps, the Unionists wanted to punish the anti-Unionists both for their unpatriotic stance and defiance to end slavery in Maryland. In a sense, the Emancipation Declaration signed a year earlier by President Lincoln might have provided the boost that the anti-slavery crusaders needed to abolish slavery in Maryland. After a much rancorous debate about the issue of slavery, the convention delegates, mainly the Unionists majority, prevailed in making slavery illegal by asserting that "all persons held to service or labor as slaves, are hereby declared free."[20]

Three years later, Democratic legislators in Annapolis called for another constitutional convention. This particular Constitution "repealed the iron-clad oath, and while rejecting blacks as voters, gave blacks the right to testify in court proceedings."[21] Although Maryland did not grant blacks the right to vote in the 1867 Constitution, the Fourteenth Amendment of the U.S. Constitution granted blacks the right to vote even in Maryland elections.[22]

Somerset and Worcester counties were initially opposed to the adoption of the 1864 Constitution. They wanted to maintain the status quo, which included legalized slavery as well as second-class citizenship for those blacks who were free. However, the counties reluctantly acquiesced, as they were unable to stall the adoption of the Constitution. Other Maryland counties provided the votes that were needed to adopt the constitution.[23] In 1867, Lt. Governor Christopher C. Cox, the presiding officer of the Maryland State Senate, echoed the sentiments of Lower Shore white citizens when he proclaimed that Maryland was not prepared for social equality of the races.[24]

In a sense, Maryland had limited option in deciding whether to free the slaves or continue the practice. States like Texas, Arkansas, Mississippi, Georgia, North and South Carolina, Louisiana, Tennessee, and Alabama contemplated compliance with President Abraham Lincoln's Emancipation

Proclamation. This meant that Maryland, along with Kentucky and Missouri, would be the only remaining states to free their slaves. It seemed then that Maryland had no choice but to free its slaves—not out of a moral conviction of the inhumanity of slavery, but rather out of necessity.[25]

After the Civil War, although many African Americans left their lands for towns and cities, others stayed and continued to farm in the areas where they had lived before 1865. Diversification led to the cultivation of other crops such as vegetables and fruits. It also brought about the harvesting and processing of oysters and crabs and the processing and canning of tomatoes and other vegetables, which dated back to the latter part of the 1800s when two white men operated a cannery in Pocomoke, Worcester County.[26]

CREATION OF THE LOWER EASTERN SHORE COUNTIES

The creation of Lower Maryland Eastern Shore counties is steeped in American revolutionary history. This region and the state were active participants in the American Revolution in the sense that there were strong sentiments of pro and anti-British feelings. The pro and anti revolutionary sentiments were played out on the Lower Maryland Eastern Shore beginning in 1666 when the first of many attempts to populate the boundaries beyond the Chesapeake Bay began in Somerset. The [English] Provincial Council issued an order that established Somerset County and named it in honor of Lady Somersett, who was the sister of Lady Arundell, wife of Cecilius Calvert, Second Lord Baltimore. The history of the county is steeped in defiance. The county came about after much discontent between the community of Quakers and the Calvert family over religion. The Quakers refused to conform to the religious values of the Calverts and Roman Catholicism. As a consequence, Somerset County was formed out of defiance, which had been its defining characteristic. According to the historian, John R. Wennersten,

> In those years the royal colony of Virginia embarked on a policy of religious persecution of the large community of Quakers then resident in Accomack County on Virginia's Eastern Shore. Rather than submit to the domination of crown and established church, the Quakers simply packed up and left . . . at that time, [Quakers] were [viewed as] unreasonable and turbulent sort of people . . . teaching lies, miracles, false visions, prophesies, and doctrines tending to disturb the peace, disorganize society, and destroy all law and government and religion.[27]

The Quakers' resoluteness laid the foundation for the county's reputation as independent-minded. Throughout much of Maryland's history, Somerset

County has held the reputation as a county that prides itself as capable of self-reliance and succeeding against all odds. Such independent-mindedness has been a driving force in shaping the relations between the seat of government in Annapolis and the region as was evidenced by the vocal assertiveness of Lower Shore citizens as well as their popular and defiant song, "We don't give a damm 'bout the whole state of Maryland.' We're from the Eastern Shore."[28]

In 1742, Worcester County was created out of Somerset County by an Act of the Maryland Legislature in honor of the Earl of Worcester [England], at the time a loyal supporter of the Crown [royal House of Stuart].[29] The principal industries of the County "are agriculture, manufacturing of lumber . . . oyster and other fisheries."[30]

In the early period, agriculture was the mainstay of the county's economy. Furthermore, the county's proximity to the Atlantic Ocean during the heyday of slave importation made it an attractive disembarking point for blacks who were brought into the state to work on the plantations. However, in 1785, the citizens of Worcester County sent a petition to the Maryland House of Delegates and asked that slavery be abolished in the county. The citizens asked that slavery be abolished largely because its practice was incompatible when juxtaposed with America's ideals of individual freedoms, rights, and human dignity.[31] Antislavery sentiments were prominent in Worcester County in part because tobacco was no longer the dominant crop after the Revolution. Slaves then were viewed more as burdens rather than assets. Their presence in significant numbers no longer held the same significance as less slave labor was needed to do other agricultural work such as grain cultivation.[32]

Although antislavery thought existed in the county, it did not prevent some whites from owning slaves for the agricultural labor that was needed on the plantations. According to the first U.S. Census conducted in 1790, Worcester County was recorded as having 7,676 whites, 3,836 slaves, and 178 free blacks, and local court records also showed that Somerset County had a free black population of "93 individuals, only 48 of whom were adults."[33] In 1800, Worcester County had 446 free blacks, and by 1860, there were 3,500 free blacks.[34]

While free blacks enjoyed a relatively "better" status than black slaves, true freedom was elusive as they formed their own communities apart from the white and slave communities.[35] The free black communities with distinct names such as "Freetown," motivated and inspired the slave communities with hopes of freedom. In spite of having their own communities, free blacks still had to contend with white hostilities because their freedom came at the expense of the white slave merchant-planters. The hostility continued in all aspects of life for blacks on the Shore. The free blacks in Worcester County in particular were active participants in the Maryland Colonization Society, which sought to help freed slaves wishing to leave the United States and move to Africa (Liberia) for permanent settlement.[36]

Wicomico County was carved out of Somerset and Worcester Counties in 1867 and named after Wicomico (wicko mekee) Indians who lived by the Wicomico River. The county was created after much debate which raged between Somerset and Worcester Counties over the commercial and port city of Salisbury as well as in response to the sustained demand by the natives of Salisbury for the creation of the new county. Natives of both Somerset and Worcester were locked in a debate about how to fairly share the commercial benefits of the prosperous city.[37] In the end, the creation of Wicomico was a sort of a compromise between Somerset and Worcester counties to accommodate Salisbury natives. Since its creation, the county grew in population and became the most populated of all the counties on the Maryland Eastern Shore.[38]

The attention that Salisbury received brought with it much discussion and interest about its regional and strategic role in preserving the regional commercial interests of the entire Eastern Shore. The context of the city's significance lies in the fact that it was chartered in 1732 and was in existence long before Wicomico and Worcester Counties were formed. The residents of Salisbury were proud of their anti-Union stance and the city's maritime commerce reaped much benefit for Somerset and Worcester counties. Salisbury natives soon began to clamor for their own county separate and apart from Somerset and Worcester counties because of bitterness and resentment that resulted from the way county officials conducted or failed to conduct business as some radical elements within the Salisbury community expected.[39] Wicomico natives and the rest of Maryland Eastern Shore had southern-like values and prided themselves in individualism and independence. It insulted their pride to have Union soldiers present on the Shore. Not only was their presence an affront, Union soldiers used Salisbury as a base to counter the strong anti-Union and pro-Confederate sentiments.[40]

Somerset and Worcester counties, to these radical elements, still represented and reminded them of British presence and influence in their midst. After all, both counties were named in honor of English prominent and influential figures who were affiliated with the Crown. For native Salisburians, it seemed that leaving the economically flourishing Salisbury in the hands of Somerset and Worcester county officials would have been a betrayal of local patriotism. In light of these considerations, Salisbury natives petitioned the Maryland Legislature meeting at Annapolis where they formed alliances with sympathetic elements in the legislature and successfully prevailed, leading to the creation of a new county in 1867.[41] In the decades following its creation,

Salisbury, and by extension Wicomico County, has been the commercial engine of the region and played an important role in both the early as well as the large-scale phases of commercial poultry production in Maryland.

SOCIAL SETTING

In the nineteenth and twentieth centuries, the Lower Maryland Eastern Shore was a tense and racialized region and relations between whites and blacks were precarious. It was in this region that the last black lynchings in Maryland were reported. In 1931, an African American, Euel Lee, also known as Orphan Jones and accused of murdering a white family in Berlin, Worcester County, was lynched by a white mob in Salisbury, Wicomico County.[42] In the same year, the body of another African American, Matthew Williams, who allegedly shot a white man and then shot himself to death, was mutilated and dragged through the street and then set on fire. Two years later, an African American, George Armwood, who was alleged to have "raped" a white woman, was lynched in Princess Anne, Somerset County.[43]

These incidents demonstrated the volatility of the Lower Eastern Shore. The lynchings also indicate that life for African Americans was fraught with persistent violence and fear. It was in this environment that blacks lived and survived. Also, local white officials dominated all aspects of life and treated blacks as second-class citizens "with no rights that whites were bound to uphold and respect."[44] Their domination relegated blacks to the periphery of economic life. That is, blacks were not in positions to dictate or influence the direction of the local economy. Instead, their influence in the local economy was confined to being the labor underclass.

Life on the Lower Maryland Eastern Shore was rural centered. Blacks worked mainly in agriculture. In terms of education, segregation was the norm of regional life until the second half of the twentieth century. In particular, public education was segregated from its beginnings in the 1870s until well after the *Brown v. Board of Education* decision in 1954.[45] On the Lower Eastern Shore and contrary to local pretensions, educational opportunities for blacks and whites were woefully unequal. Black students received and used outdated books passed down from the better-funded white schools. The attendant consequence of such practice was that it placed black students at a disadvantage. Without an education to overcome this disadvantage, blacks were more likely to end up as the labor underclass of commercial poultry production.

Although desegregation of public education in all three counties occurred after the *Brown* decision, Lower Shore county officials moved at a much slower pace than officials in Baltimore and other jurisdictions of Maryland. More than a decade after the 1954 *Brown* decision and when other parts of the state were engaged in ways to implement the *Brown* decision without compromising their racial parochialisms, the Lower Shore counties dragged their feet on desegregating the schools.[46]

Segregation in public education was not restricted to K–12. It continued all the way through higher education until the court-ordered desegregation of schools following the precedent set by the Brown decision. The Brown decision forced the University of Maryland's flagship campus—College Park and other formally all-white state colleges to lift their previous ban on admitting blacks into the colleges. Meanwhile, historically black colleges including Bowie, Coppin, and Morgan State Colleges and the Maryland State College, which later became the University of Maryland Eastern Shore (UMES) were the training grounds for blacks in Maryland. The State's flagship university in College Park had its own history of racism, segregation, and discrimination. For example, its law school remained segregated until 1935 when it was forced by a court order to grant admission to a black student, Donald G. Murray.[47] Its undergraduate programs, however, remained segregated until the 1950s.[48]

Moreover in 1888, the Maryland legislature named such land-grant universities as UMES and College Park as the state agricultural experiment stations[49] to develop ways to address the needs of farmers including blacks in mainly rural and agrarian areas and teach them skills to enhance their farming lifestyle.[50] The objective of the stations was to make a connection between farmers and experiment station researchers. The first black extension service agent (they were also called county agents) was appointed in 1917. His task was to liaise with blacks in Maryland including Somerset, Wicomico, and Worcester counties about their farming. This agent became, in most cases, the only link between black farmers and the government.[51]

The first black extension agents for the Lower Shore and southern Maryland were school teachers.[52] These black teachers were affiliated with the Princess Anne Academy. The state recognized that black teachers commanded respect and were highly regarded in the black community. Many blacks identified with the agricultural extension agents because teachers of all sorts were instrumental in black collective uplift, pride, and identity formation.[53]

ECONOMIC SETTING

Although Maryland welcomed those yearning to be free, it became one of the strongest advocates and practitioners of slavery, using slave labor to build its tobacco economy. In early Maryland's history, Africans came as indentured servants. After their indenture, they became farmers and some became landowners.[54] Their presence in Maryland, home to tobacco, a major export crop in the early colony, was critical to the colony's economic viability. For example, the topography and the vegetation of southern Maryland where many of the early blacks settled provided impetus for blacks' early involvement in agriculture. In other areas of the state, agriculture was a way of life. The Lower Eastern Shore in particular developed a stable economy based on agriculture.[55]

On the Lower Eastern Shore, the soil was more conducive for cultivating fruits and vegetables because of good soil. As noted by Donald Marquand Dozer, agriculture in this area flourished. According to Dozer, farmers benefited from:

> . . . , abundant rainfall, and a long growing season extending from early April to the end of October. It is one of the most productive farming sections of the United States for small fruits and vegetables. In the early years of tobacco culture the Eastern Shore specialized in this crop, but its tobacco production gradually declined and disappeared entirely because of soil exhaustion, poor marketing conditions, and priorities given to fruit and truck crops. These latter agricultural specialties of the Eastern Shore include strawberries, peaches, pears, asparagus, corn, cucumbers, watermelons, cantaloupes, peas, beans, white and sweet potatoes, wheat, and tomatoes. Here the canning and freezing of these crops is a main industry. Dairying and poultry farming are also big business on the Eastern Shore and the cutting of timber, particularly pine, and the manufacture of forest products provide a livelihood for many residents on the lower Eastern Shore counties.[56]

Blacks were important participants in agriculture. Although most African Americans were indentured servants, their status later changed to slaves around 1661 after the colonial legislature enacted laws that made slavery legal.[57] As the legal status of blacks in Maryland changed, the newly arrived Africans came as enslaved persons, bound for life. Blacks then bore the burden of providing the labor for Maryland's agriculture.[58] In essence, African Americans have been the labor underclass on the Lower Maryland Eastern Shore since its settlement. Their involvement in the labor supply-side of agricultural production including commercial poultry production was not a twentieth century phenomenon.

By the mid-twentieth century, there was agricultural diversification. Commercial poultry production became a major component of the county's economy with the establishment of poultry processing plants in Berlin in 1940 and another plant in Pocomoke in 1942.[59] African Americans played an important role on the labor supply-side of commercial poultry production because a vast majority of the processing plant workers were African Americans.[60]

Also in the 20th century, there was a correlation between population and agriculture. Poultry became a major agricultural product accounting for more than half of all agricultural products from Somerset, Wicomico, and Worcester counties.[61] By 1930 and 1940, Maryland in general experienced a steady increase in its population. For example, between 1950 and 1990, the population of the Wicomico County grew by more than 80 percent, and the African American population grew by nearly 100 percent from 8,372 to 16,573.[62] Most blacks were drawn to Salisbury, home of Perdue Farms, by the presence of poultry work and its need for an unskilled labor force. In this period, the state also experienced an increase in black population with the exception of 1950 when the black population decreased by 0.8%.[63]

The census figures for the selected years in Tables 2.1–2.4 establish the context in which this work takes place: the black population constituted the

Table 2.1. Maryland Census, 1930–1990

Year	*Population*
1930	1,631,526
1940	1,821,244
1950	2,343,001
1960	3,110,689
1970	3,922,399
1980	4,216,975
1990	4,781,468

Source: U.S. Bureau of the Census of Population.

Table 2.2. Somerset County Population, 1930–1990

	1930	*1940*	*1950*	*1960*	*1970*	*1980*	*1990*
Total	23,382	20,965	20,745	19,623	18,924	19,188	23,440
White	15,271	13,904	13,904	12,263	11,800	12,433	14,282
%	65.3%	66.3%	67.0%	62.4%	62.4%	64.8%	60.9%
Black	8,111	7,061	7,326	7,360	7,113	6,639	8,943
%	34.6%	33.6%	35.3%	37.5%	37.6%	34.6%	38.2%
Hispanic+	NR	NR	NR	NR	NR	125	229
%	—	—	—	—	—	0.7%	1%

NR = Not Reported

+Maryland Census did not capture Hispanic populations in the Lower Shore counties until 1970. In that year, the Hispanic population was reported only for counties with a minimum of 400 persons speaking a Spanish language. Perhaps the unwillingness of some Hispanics to be counted at all meant that the correct figures were probably higher.

Source: U.S. Bureau of the Census; Maryland Historical Census.

Table 2.3. Wicomico County Population, 1930–1990

	1930	*1940*	*1950*	*1960*	*1970*	*1980*	*1990*
Total	31,229	34,530	39,641	49,050	54,236	64,540	74,339
White	24,471	27,035	30,795	38,026	42,636	49,679	56,755
%	78.3%	78.2%	77.6%	77.5%	78.6%	77.4%	76.3%
Black	6,750	7,477	8,372	11,024	11,512	14,085	16,573
%	21.6%	21.6%	21.1%	22.4%	21.2%	21.8%	22.3%
Hispanic	NR	NR	NR	NR	443	408	610
%	—	—	—	—	0.8%	0.6%	0.8%

NR = Not Reported

Source: U.S. Bureau of the Census; Maryland Historical Census.

Table 2.4. Worcester County Population, 1930–1990

	1930	*1940*	*1950*	*1960*	*1970*	*1980*	*1990*
Total	21,624	21,245	23,148	23,733	24,442	30,889	35,028
White	14,912	14,575	16,046	15,585	16,397	22,593	27,253
%	68.9%	68.6%	69.3%	65.6%	67.1%	73.1%	77.8%
Black	6,712	6,669	7,094	8,148	8,017	8,100	7,467
%	31.0%	31.3%	30.6%	34.3%	32.8%	26.2%	21.3%
Hispanic	NR	NR	NR	NR	NR	235	275
%	—	—	—	—	—	0.8%	0.8%

NR = Not Reported

Source: U.S. Bureau of the Census; Maryland Historical Census.

labor underclass of commercial poultry production on the Lower Maryland Eastern Shore. In all three counties, African Americans made up a significant part of the population accounting for over one third of the total population. Consequently, blacks have played an important role in the labor supply-side of commercial poultry production.

In Wicomico County, the size of the black population grew faster than that of the overall population, perhaps due to the economic prospects of the burgeoning commercial poultry industry. The increased population in Wicomico County also provided the industry with available labor pool from which to draw. Likewise in Worcester County, an area dominated by the seashore town of Ocean City, poultry growing is a significant industry and many, mostly whites, raise poultry for large-scale commercial poultry companies such as Perdue Farms, Tyson Foods, and Mountaire Farms.

Food production and agrarianism were major economic aspects of black life on the Lower Shore. According to the *Agricultural Year Book*, a yearly publication that tracks the state agricultural sector, in 1920, there were 3,549 farms operated by black owners and 2,509 farms operated by black tenants and croppers in Maryland.[64] The number of black farm owners prior to the great depression is instructive because either their families or white benefactors bequeathed many of the farms to them or in turn many of the black farmers remained on the farms and ensured that the tradition continued.[65]

The Eastern Shore counties are also the breadbasket of the state. A large part of Maryland's agriculture occurs on the Lower Eastern Shore. Specifically, more than two thirds of Maryland's commercial poultry production takes place on the Lower Shore (Somerset, Wicomico, and Worcester Counties).[66] Given the fact that poultry is widely consumed in Maryland, the region is of a strategic and economic importance to the rest of the state. Therefore, both the region and its people are relevant to the state because they produce important economic products.

The rural/urban population distribution indicates the importance of agrarian life-style in the counties, which made the counties to be mostly rural. The arability of the land for agricultural farming (typically, farming areas are usually rural) accounts for the predominance of rural life-style. However, Wicomico County was the most urban of all three counties in part because of the location of Salisbury, an important port city and job center for the region. Somerset County was and has remained the most rural. Although the state has remained predominantly urban since 1910, the Lower Eastern Shore has remained rural. The rural setting, with much rich soil and good poultry growing, gave the region a unique suitability for meeting the dietary and nutritional needs of Marylanders.

Since farming dominated the rural lifestyle, most African Americans had few options besides farming. The residual effects of segregation and discrimination are still evident as blacks and whites still live in their own enclaves. An unwritten understanding exists whereby blacks only sought particular types of jobs. Even blacks with college-level education were not assured front-office positions in the corporate offices of the poultry companies. A local resident narrated the story of a family member who worked in the processing plant as a line worker and attended one of the local universities while working in the plant. She graduated with a bachelor's degree but was denied a front-office position in the corporate office of one of the local major poultry companies. Another local resident noted that "there were instances where some blacks who worked and still work in the plants already had degrees but could not work elsewhere because there were few to no other opportunities. I have friends and relatives who are college graduates who have worked in the plant for a long time and still work there."[67] However, because of this sort of job discrimination, some workers have chosen to leave the region and seek better jobs elsewhere.

Perhaps blacks would not have dominated the labor underclass of the poultry industry had discrimination, racism, and segregation not been the social norms of the Lower Maryland Eastern Shore. For example, contract farming is an important aspect of commercial poultry production. Under a contract arrangement, with the most common type in the 1950s and 1960s being the flat-fee contracts, the poultry companies offered contracts to growers and provided the chicks, feed, and medication.[68]

The companies provided everything except the poultry house, which was the responsibility of the growers. These houses had to meet the standards and specifications set by the companies. The growers also agreed to purchase their poultry feed and other materials from the companies. The growers had the responsibility to dispose the dead chicks. Many blacks could not become contract growers in part because they lacked the financial resources and collateral. Also, if the credit and lending requirements as adopted by the poultry companies had been liberally extended, perhaps more blacks could have become growers and hired their own crew to catch chickens in the poultry houses. Despite the racial considerations and realities, class, nevertheless, triumphed race as the white underclass could not secure growers' contracts or even own poultry houses. Ownership translated to economic power on the Lower Shore. Those who owned poultry facility or facilities were part of the higher economic class.

The census figures in Tables 2.5–2.7 give a concise overview of population in Maryland and the Lower Maryland Eastern Shore. Particularly, the white and black population distribution speaks to the inter-dependence between the two races—although whites dominated the managerial and supervisory positions while blacks provided the labor.

While there was ruralism of many southern states during the early to middle part of the twentieth century, blacks out-migrated from the South to the urban areas in the mid-West and the North such as Chicago and New York City. Maryland by the 1930s was an urban state. Nearly 60% of Marylanders lived in urban areas. In 1930, of the 40.2% of Marylanders who lived in the rural areas, 11.6% lived on the Lower Maryland Eastern Shore.

In 1930, the Somerset County population was predominantly white with a sizeable black population. The county was also overwhelmingly rural with more than 80% of the population living in the rural areas compared to 16.4% who lived in the urban areas. In the same year, 35.2% of Wicomico County residents were urban dwellers while only 12.0% of Worcester County residents were urban dwellers. Between 1950 and 1990, the African American population increased in all three counties from 35.3 percent to 38.2 percent.[69]

The three counties combined averaged 76.6% rural population compared to 40.7% statewide in 1940. In 1950, the average rural population in the three counties (76.7%) remained virtually the same as that of 1940, while the state's rural population declined to 31.0%. Although there was a decline

Table 2.5. Maryland Population by Race for Selected Years, 1930–1990

	1930	*1940*	*1950*	*1960*	*1970*	*1980*	*1990*
Total	1,631,526	1,821,244	2,343,001	3,100,689	3,922,399	4,216,975	4,780,743
Whites	1,354,226	1,518,481	1,954,975	2,573,919	3,193,021	3,158,838	3,393,964
%	83.5%	82.6%	83.4%	83.0%	81.4%	74.9%	71.0%
Blacks	276,379	301,931	385,972	518,410	701,341	958,150	1,189,899
%	16.4%	17.3%	16.5%	16.7%	17.9%	22.7%	24.9%
Hispanic	NR	NR	NR	NR	52,974	64,746	125,102
%	—	—	—	—	1.3%	1.5%	2.62%

*The statewide figures for Hispanics did not start showing up on the Census records for Maryland until 1970. By this time, the category was generic and comprised of Indians, Chinese, and Japanese. Non-natives were classified in the "other races" category along with Spanish speaking people.

Source: U.S. Bureau of the Census; Maryland Historical Census.

Table 2.6. Maryland Urban/Rural Population Distribution for Selected Years, 1930–1990

	1930	*1940*	*1950*	*1960*	*1970*	*1980*	*1990*
Urban	974,869	1,080,351	1,615,902	2,253,832	3,003,935	3,386,693	3,888,429
%	59.8%	59.3%	69.0%	72.7%	76.6%	80.3%	77.1%
Rural	656,657	740,893	727,099	846,857	918,464	830,282	893,039
%	40.2%	40.7%	31.0%	27.3%	23.4%	19.7%	22.9%

Source: Maryland Historical Census.

Table 2.7. Urban and Rural Population Distributions for the Three Lower Maryland Eastern Shore Counties, 1930–1990

	State	*Somerset*	*Wicomico*	*Worcester*
1930				
Total	1,631,526	23,382	31,229	21,624
Urban	974,869	3,850	10,997	2,609
%	59.8%	16.4%	35.2%	12.0%
Rural	656,657	19,532	20,232	19,015
%	40.2%	83.5%	64.7%	87.9%
1940				
Total	1,821,244	20,965	34,641	21,245
Urban	1,080,354	3,908	13,313	2,739
%	59.3%	18.6%	38.6%	12.9%
Rural	740,893	17,057	21,217	18,506
%	40.7%	81.4%	61.4%	87.1%
1950				
Total	2,343,001	20,745	39,641	23,148
Urban	1,615,902	3,688	15,141	3,191
%	69.0%	17.7%	38.2%	13.8%
Rural	727,099	17,057	24,500	19,957
%	31.0%	82.2%	61.8%	86.2%
1960				
Total	3,100,689	19,623	49,050	23,733
Urban	2,253,832	3,540	16,302	3,329
%	72.7%	18.0%	33.2%	14.0%
Rural	846,857	16,083	32,748	20,404
%	27.3%	82.0%	66.8%	86.0%
1970				
Total	3,922,399	18,924	54,236	24,442
Urban	3,003,935	3,078	15,252	3,573
%	76.6%	16.3%	28.1%	14.6%
Rural	918,464	15,846	38,984	20,869
%	23.4%	83.7%	71.9%	85.4%
1980				
Total	4,216,975	19,188	64,540	30,889
Urban	3,386,693	2,924	19,123	8,504
%	80.3%	15.2%	29.6%	27.5%
Rural	830,282	16,262	45,417	22,385
%	19.7%	84.8%	70.4%	72.5%
1990				
Total	4,780,743	23,440	74,339	35,028
Urban	3,888,429	2,880	24,103	15,903
%	81.3%	12.29%	32.42%	45.40%
Rural	893,039	20,560	50,236	19,125
%	18.7%	87.1%	67.58%	54.60%

Source: U.S. Census of Population; State of Maryland Department of Planning, [Historical Census, 1930-1990].

in statewide rural population, the people that migrated did not relocate to the Lower Eastern Shore, perhaps in part due to the predominance of agriculture.

By 1960, the statewide rural population declined again to 27.3% while the combined rural population for the three counties increased to an average of 78.2%. The increase in population might have been as a result of intra-migration rather than people migrating into the counties from other parts of the state. Also, the birthrate was another probability that could likely explain the slight increase. Nevertheless, the region remained rural in part because of socio-economic factors such as low education and poverty. For example, in Somerset County, white male heads of household had an average of 8.9 years of school completed while the white female wives had 9.5 years of school completed. For blacks, male heads of household had an average of 7.2 years and black female wives had 8.5 years of school completed.[70]

The state in general became increasingly more urban than rural in 1970. The statewide rural population declined yet again to 23.4% from 27.3% a decade earlier. On their part, the three counties registered an increase in rural population from 78.2% in 1960 to 80.3% in 1970. The state was becoming urbanized; the Lower Maryland Eastern Shore was becoming more rural, indicating an agricultural tradition that was passed on from one generation to another.

In 1980, the three counties still had the highest average of rural population in Maryland although it had decreased from 80.3% in 1970 to 75.9% in 1980. The statewide rural population declined as its urban population increased. Worcester County saw the most decrease in percentage in rural population from 85.4% in 1970 to 72.5% in 1980 although the absolute number of rural residents increased.

One explanation for this rural decrease was the increase in the tourism industry that drew people to urban areas which saw a large increase in population during the decade of the 1970s. In particular, Ocean City was a major tourist attraction. Another significant reason for the shift in the rural and urban dynamics of the three counties was the re-classification and redefinition of rural and urban areas by the U.S. Census Bureau. The new classification redefined areas with 2,500 persons or more as cities. The previous classification gave the three counties their rural designation because most people lived in small townships spread throughout the counties. The declining trend in rural population continued mainly in the state at 18.7% and in Wicomico County it was 67.5%, while it was 54.6% in Worcester County in 1990.

Somerset was the only county that remained highly rural at 87.17% in 1990. For Wicomico County, the influx of Latino workers into Salisbury and the influx of economic opportunities that the city presented as a major regional commercial base hold one explanation for the decrease in its rural population. However, during the decades of the late twentieth century, it seemed that out-migration occurred in the rural counties of Somerset and Worcester to the more urban Wicomico County because of the availability of low-skilled jobs in commercial poultry production and other service industry jobs.

The distribution of rural and urban population indicates that the Lower Maryland Eastern Shore was the food and fiber region of the state. The high concentration of its people in the rural areas along with favorable weather and soil conditions enabled the three counties to focus on poultry and other food production. The statistics also reveal that poultry can best be produced in a rural setting where the production of broiler feed is localized. That is, many of the grain producers are located in the Delmarva region where much of the commercial poultry production has been vertically integrated an arrangement that has given big poultry companies a competitive edge over independent poultry producers. The convenience of having the chicken feed companies localized meant lower costs for chicken feed. This convenience also meant reduced transportation costs, some of which would have been passed on to the consumer.

CONCLUSION

The Lower Maryland Eastern Shore counties have been an important region of Maryland. For example, Somerset County, one of the "original"[71] counties, was the first county on the Lower Shore to be established in 1666. The region's participation in the economic wellbeing of the state has been noteworthy. Besides commercial poultry production, the three Lower Shore counties have also been important producers of other agricultural products such as tobacco, fruits, and vegetables. However, poultry became the dominant agricultural product soon after Frank Perdue, a local Shoreman, raised his first flock that began and ultimately transformed commercial poultry production in Maryland. The topography of the region has also made it conducive to raise poultry.

With regard to African Americans, they have been in Maryland since its founding in 1634 and on the Lower Eastern Shore throughout the antebellum period. Their presence has not been without much oppression and violence. Racial relations were particularly tense even through the civil rights movement and beyond. In the face of such tension, local whites enforced strict segregation and discriminatory laws to assert their power and dominance. Subsequently, blacks' presence on the Lower Maryland Eastern Shore then has been mainly as the labor underclass whether on the tobacco plantations, fruit and vegetable fields or in poultry production. In a sense, black labor was essential to the Lower Maryland Eastern Shore's economy and by extension to the entire state since poultry accounted for more than a quarter of the state's agricultural income.[72] Their presence proved beneficial at the early phase of the commercialization of poultry production in Maryland and remained so with the increased commercialization of the industry.

NOTES

1. James C. Mullikin, "The Eastern Shore," in Morris L. Radoff, (ed.,) *The Old Line State: A History of Maryland* (Annapolis, Maryland: Hall of Records Commission, State of Maryland, 1971), 149–161.
2. Vincent Wilson, Jr., *The Book of the States* (Brookville, Maryland: American History Research Associates, 1992), 46.
3. [Maryland: Land and Area] Maryland State Archives. Available at http://www.mdarchives.state.md.us/mdmanual and Maryland Geological Survey.
4. George R. Shivers, *Changing Times: Chronicle of Allen, Maryland, An Eastern Shore Village*. (Baltimore: Gateway Press Inc., 1998), 5–10.
5. Susan Ellery Greene Chappelle, Jean H. Baker, Dean R. Esslinger, Whitman H. Ridgway, Jean B. Russo, Constance B. Schulz, and Gregory A. Stiverson, *Maryland: A History of Its People* (Baltimore: The Johns Hopkins University Press, 1986), 201.
6. J. Montgomery Gambrill, *Leading Events of Maryland History With Topical Analyses, References, and Questions for Original Thought and Research* (Boston: Ginn and Company, n.d.), 249 [Maryland Room, Enoch Pratt Free Library, Baltimore].
7. Chappelle, 201.
8. Glenn O. Phillips, "Maryland and the Caribbean, 1634–1984: Some Highlights," *Maryland Historical Magazine*, Vol. 83, No. 3 (Fall 1988), 199–214.
9. Matthew Page Andrews, *History of Maryland: Province and State* (Hatsboro, Pennsylvania: Tradition Press, 1965).
10. See Harry Wright Newman, *The Flowering of the Palatinate* (Washington, DC,: [n. p.] 1961), 38; [reprinted, 1998], Baltimore, Maryland: Genealogical Publishing Co., Inc., and Alice Norris Parran, Series II, "Register of Maryland's Heraldic Families" [n.p.] (1938), 339–343 [Maryland Room, Enoch Pratt Free Library, Baltimore].
11. See John R. Wennersten, "A Cycle of Race Relations on Maryland's Eastern Shore: Somerset County, 1850–1917," *Maryland Historical Magazine*, Volume 80, No. 4 (Winter 1985), 377–382.
12. Donald D. Max, "Black Immigrants: The Slave Trade in Colonial Maryland," *Maryland Historical Magazine*, Volume 73, No. 1 (March 1978), 31–45.
13. Margaret L. Andersen, "Discovering the Past/Considering the Future: Lessons from the Eastern Shore," in Carole C. Marks, (ed.), *The History of African Americans of Delaware and Maryland's Eastern Shore* (Wilmington, Delaware: Delaware

Heritage Commission, 1998), 106. See also, Richard Walsh and William Lloyd, *Maryland: A History, 1632–1974* (Baltimore: Maryland Historical Society, 1974).

14. Barry Neville and Edward Jones, "Slavery in Worcester County, Maryland, 1688–1766," *Maryland Historical Magazine*, Volume 89, No. 3 (Fall 1994), 319–327.

15. "Didn't It Rain: Civil War," *African American Heritage in Worcester County* [brochure, n.d], published by the Worcester County African American Heritage Society and gives a historical timeline of African Americans in Worcester County using real but selected pictures and artistic renditions to illuminate black life in the county.

16. See Margaret L. Andersen, in Carole C. Marks, (ed.), 106.

17. George H. Corddry, *Wicomico County History*. (Salisbury: Peninsula Press, 1981), 55.

18. Maryland State Constitution, 1864 [Maryland State Archives].

19. Robert J. Brugger, *Maryland: A Middle Temperament, 1634–1980* (Baltimore: The Johns Hopkins University Press, 1988), 302–305.

20. Ibid, 304.

21. Chappelle, 172

22. Brugger, 306.

23. Chappelle et al., *Maryland: . . .* , 135.

24. Corddry, 55. See also a photographic image of Lt. Governor Cox at Maryland State Archives, Special Collections, and [MSA SC 2214-1-31].

25. "Slavery Practically Abolished," *Harper's Weekly*, 4 October 1862 [Maryland Room, Enoch Pratt Free Library, Baltimore]. See also, Richard Paul Fuke, *Imperfect Equality: African Americans and the Confines of White Racial Attitudes in Post-Emancipation Maryland* (New York: Fordham University Press, 1999).

26. See R. Lee Burton Jr., *Canneries of the Eastern Shore* (Centerville: Maryland: Tidewater Publishers, 1986) for an extended treatment of this industry on the Lower Shore.

27. Wennersten, 90.

28. Ibid, 3.

29. Gambrill, 249.

30. Ibid, 250.

31. Wennersten, 121. See also, Clara L. Small, "Abolitionists, Free Blacks, and Runaway Slaves: Surviving Slavery on Maryland's Eastern Shore," in Carole C. Marks, (ed.,), *History of African Americans of Delaware and Maryland's Eastern Shore* (Wilmington, Delaware: Delaware Heritage Commission, 1998), 55–73.

32. Ibid.

33. Worcester County MDGenWeb Project [Worcester History—Historic Timeline of Worcester County]. Available at http://www.rootsweb.com/~mdworces/history.htm See also, Thomas E. Davidson, "Free Blacks in Old Somerset County, 1745–1755," *Maryland Historical Magazine*, Volume 80, No. 2 (Summer 1985), 151–156.

34. "Go Down Moses: Free Blacks," *African American Heritage in Worcester County*, [brochure, n.d.].

35. I am using "better" here in a restrictive sense. Free blacks were in a better position than slaves because they did not have to maintain the same rigid work schedule

as other slaves such as tilling the fields. Also, free blacks were in a better position because they worked for themselves and had a measure of economic independence.

36. *African American Heritage in Worcester County,* [brochure, n.d]. See also, Ross M. Kimmel, "Free Blacks in Seventeenth Century Maryland," *Maryland Historical Magazine*, Volume 71, No. 1 (Spring 1976), 19–25.

37. Wennersten, 75.

38. See Population tables, 1930–1990.

39. Wennersten, 75.

40. Ibid, 74.

41. Ibid.

42. For a more extended treatment of this case, see Joseph E. Moore, *Murder on Maryland's Eastern Shore: Race, Politics and the Case of Orphan Jones* (Charleston, South Carolina: History Press, 2006).

43. See Cezar Tampoya Jackson, "A Comparative Study of Perceptions of the Media Relating to Lynchings on the Eastern Shore of Maryland, 1931–1933" (M.A. thesis, Salisbury State University, 1996) for a more extended discussion of the case. Though the facts of the case as reported by the media pointed to a simple robbery, soon the word rape was employed by whites who were bent on teaching black people a lesson about respecting white womanhood. The white mob used "rape" to evoke racial sentiments and mobilize other whites to supposedly protect one of their own. In actuality, the white victim was not raped. Instead, she was robbed. Another important and relevant work by Sherrilyn Ifill, *On the Courthouse Lawn: Confronting the Legacy of Lynching in the 21st Century* (Boston: Beacon Press, 2007), explores historical lynchings on the Eastern Shore.

44. See *Dred Scott v. Sanford, 60 U.S. 393* (1856); A case to which many southern whites pointed in justifying their treatment of blacks. Although the case was heard and decided by the U.S. Supreme Court nearly a century earlier, whites' attitudes towards blacks still lacked respect.

45. See for example, *Brown v. Board of Education*, 347 U.S. 483 (1954) (USSC+). See also, Gary Orfield and Susan E. Eaton, *Dismantling Desegregation: The Quiet Reversal of Brown v. Board of Education* (New York: The New Press, 1996), and Hansel Burley, "Separate and Unequal," *American School Board Journal*, Vol. 188, No. 6 (June 2001). Taunya Lovell Banks, "Brown at 50: Reconstructing Brown's Promise," *Washburn Law Journal*, Volume 44, No. 1 (Fall 2004), 31–64, and John R. Wennersten, "The Legacy of Brown v. Board of Education and Maryland's Eastern Shore." http://www.skipjack.net/article.

46. Deborah Tulani Salahu-Din, "The Forgotten Pawns," *The Baltimore Sun*, 4 May 2004.

47. Mike Bowler, "Black Students Sent Away," *The Baltimore Sun*, [op-ed] 16 May 2004. One fact that has often been omitted about the University of Maryland Law School and its admission policies of non-whites is that two African Americans graduated from the law school in 1889 before it was closed to all black students. See David S. Bogen, "The Transformation of the Fourteenth Amendment: Reflections from the Admission of Maryland's First Black Attorneys," *Maryland Law Review,* Vol.44 (Summer 1985).

48. Edward J. Kuebler, "The Desegregation of the University of Maryland," *Maryland Historical Magazine*, Volume 71, No. 1 (Spring 1976), 37–49.

49. Norwood Allen Kerr, *The Legacy: A Centennial History of the State Agricultural Experiment Stations, 1887–1987* (Columbia, Missouri: Missouri Agricultural Experiment Station), 25. See also, George H. Callcott, *History of the University of Maryland* (Baltimore: Maryland Historical Society, 1966), 189–190; A.C. True and V.A. Clark, *Agricultural Experiment Stations in the United States* (Washington, D.C.: USDA, Office of Experiment Stations Bulletin 80, 1900), 164–167, 171–174, and John R. and Ruth Ellen Wennersten, "Separate and Unequal: The Evolution of a Black Land Grant College in Maryland, 1890–1930," *Maryland Historical Magazine*, Volume 72, No. 1 (Spring 1977), 110–117.

50. The University of Maryland, Eastern Shore was founded in 1886 by the Delaware Conference for Colored Methodists as the Princess Anne Academy. In 1919, the name was changed to the Eastern Shore Branch of the Maryland Agricultural College. In 1948, it was renamed the Maryland State College. Finally, in 1970, it became the University of Maryland, Eastern Shore. The Extension Service was established in 1890 as part of the Land-Grant program for historical black colleges and universities.

51. A.B. Hamilton and C.K. McGee, "The Economic and Social Status of Rural Negro Families in Maryland." (College Park, Maryland: The University of Maryland Agricultural Experiment Station and Extension Service, 1948): 6. [*Bulletin X4*].

52. Hamilton and McGee, 7.

53. For a more elaborate discussion on Black Uplift in the twentieth century, see Kevin K. Gaines, *Uplifting the Race: Black Leadership, Politics, and Culture in the Twentieth Century*, (Chapel Hill: The University of North Carolina Press, 1996).

54. Chappelle, 22.

55. William A. Lynk, F. Howard Forsyth, and Virginia S. Krohnfeldt, "Characteristics of Families in Poverty in a Rural Maryland County," [Scientific Article A1679] (College Park: University of Maryland, College Park; Princess Anne: University of Maryland, Eastern Shore, and Agriculture Experiment Station, June 1971), 4. This study sampled 134 families (81 black and 53 white) in Somerset County, at the time considered the poorest county in Maryland. More blacks were sampled to reflect the higher proportion of impoverished blacks than whites in the county.

56. Dozer, 3.

57. Wennersten, 118.

58. Chappelle, 25–34, passim.

59. William H. Williams, *Delmarva's Chicken Industry: 75 Years of Progress* (Georgetown, Delaware: Delmarva Poultry Industry, Inc., 1998), 31.

60. Ed Covell, interview by the author, Georgetown, Delaware.

61. *Salisbury-Wicomico Commission, 1960*, [untitled brochure on economic development in the Lower Maryland Eastern Shore region, n.d].

62. Theodore J. Davis Jr., "Socioeconomic Change: A Community in Transition," in Carole C. Marks, (ed.,), *History of African Americans of Delaware and Maryland's Eastern Shore* (Wilmington, Delaware: Delaware Heritage Commission, 1998), 206.

63. U.S. Bureau of the Census of Population, 1950, Maryland; Counties.

64. Hamilton and McGee, 7.

65. Ibid.

66. *Maryland Agricultural Statistical Service*, [Historical Summaries on Poultry and Other Crops for the selected years, 1930–1990]; United States Department of Agriculture.

67. Alison Morton, interview by author, Salisbury, Maryland.

68. See the report prepared for the Maryland Agro-Ecology Center by Robert A. Chase, Wesley L. Musser, and Bruce Gardner, *The Economic Contribution and Long-Term Sustainability of the Delmarva Poultry Industry*, (College Park, Maryland: Center for Agricultural and Natural Resource Policy, 2003), 6.

69. Theodore J. Davis Jr., "Socioeconomic Change. . . ," 206.

70. William A. Lynk, et al., 12.

71. The term "original" is used in this context to mean that unlike Wicomico and Worcester Counties, Somerset County was not carved out of any other county. Instead, it was established by a direct order issued by the English Crown.

72. See *Maryland Agricultural Statistical Service* and U.S. Census of Agriculture [USDA] for the selected years, 1930–1990.

Chapter Three

Early Phase of Commercialization of Poultry Production, 1930s to 1950s

The Great Depression of the late 1920s to the 1930s, which resulted in sharp decline in the prices of agricultural products, brought about a global agricultural depression. It brought an end to the pre-1920s favorable agricultural performance during which time farmers were accorded a place of honor in society and were regarded as "gentlemen farmers."[1] The global depression changed their status as "gentlemen farmers," as they resorted to "subsistence farming."[2] In Maryland, the agricultural sector was greatly impacted by the depression since the prices of the State's agricultural products exported to Europe were affected by price fluctuations.[3]

Despite the international situation with price fluctuations, U.S. agricultural products competed in the global market. American farmers were able to shield themselves from the uncertain global market conditions in part because of the subsidies they received from the government in cultivating and marketing their farm products. These subsidies were crucial in sustaining American farmers and, by extension, the nation that relied on farmers to feed it. The subsidies further eased the unpredictability of the global agricultural market and enabled farmers to maintain a regular supply of food for a growing nation. Commercial poultry was an important part of the agricultural economy. However, the uncertainty of the global agricultural market did not have the same impact on poultry because commercial poultry production was still in its early phase by the time of the Great Depression.

FOUNDING OF COMMERCIAL POULTRY ON THE EASTERN SHORE

Commercial poultry production became a fixture on the Lower Maryland Eastern Shore only in the 1920s. Prior to the emergence of commercial poultry operations, there were independent and small-scale poultry operators who were mainly subsistent producers. The development of commercial poultry production on the Lower Maryland Eastern Shore was facilitated by the outbreaks of diseases such as boll weevil and drought. In particular, the Dust Bowl affected agricultural crops in the 1930s and 1940s.[4] Commercial poultry therefore became a viable alternative to crops that were lost to drought and crop-infesting diseases. Farmers diversified their emphasis on crop agriculture because of the attendant famine of the 1930s.[5]

Much of the credit for the early phase of commercial poultry production on the Lower Eastern Shore was attributed to Arthur W. Perdue. A descendent of early French settlers in Maryland, he worked at Adams Express Company as a railway agent stationed in Salisbury, Maryland, where he earned $45.00 per month.[6] However, in 1919, he resigned from his railway employment, and, in 1920, he established a family-owned table-egg production poultry business in Salisbury, Wicomico County. As the table egg business gained momentum, outsiders from other parts of the East Coast came to purchase his eggs. His operations slowly expanded, and, in 1925, he built his first egg hatchery.[7] This hatchery operation positioned him to gain the upper hand over other potential competitors. The new hatchery also meant that he could produce a larger volume of consumable chickens while still supplying eggs to consumers.

By 1930, Perdue's operation was well on its way to becoming an important part of the local economy and a magnet for employment of unskilled workers on the Lower Shore. His new company's presence in the region provided opportunities for African Americans that did not previously exist. Thus, commercial poultry production on the Lower Maryland Eastern Shore began when Arthur Perdue bought and kept 50 Leghorn chickens in 1919. As a railway express agent, he observed that brokers earned much money not just from breeding chickens for egg production but also for meat consumption. He later combined his 50 chickens with his wife's 25 layers and built a chicken house. He applied the knowledge of what he gained from observing the brokers during his inter-state travels to operate his egg business. The eggs from these domesticated poultry-growing were taken to the area markets and sold at profitable prices.[8]

However, in 1923, one Cecile Steele, a white housewife from Ocean View, Sussex County, Delaware, started a small-scale commercial poultry production on the Delaware side of the Delmarva Peninsula. She raised domesticated chickens and had a regular supplier who sent her young chickens. She raised these chickens for egg production and family consumption. By 1926, Steele had accumulated enough chickens and eggs to become a leading local supplier of poultry. Although Steele was on the Delaware side of the Delmarva, her operations had a much wider impact because she attracted more brokers from other parts of the Northeast who flocked to Delaware. In contrast to Perdue, Steele's influence at the early phase of commercial poultry covered more ground and clients. Thereafter, raising poultry on the entire Delmarva Peninsula became more commercialized. By the mid to late 1930s, a shift from largely subsistence to commercial poultry production had occurred. The local historian George H Corddry wrote:

> . . . Commercial poultry flourished due to seven main reasons. The "first was a mild and temperate climate that allowed chickens to be kept in relatively cool conditions for a better yield; second, good sandy soil that provided good drainage for liquids in chicken manure, which helped control diseases; third, cheap building costs due to abundant pine forests, which were used in constructing chicken houses; fourth, proximity to the markets, which eased transportation of poultry products to other parts of the country as the region included Salisbury, the second biggest and most important seaport in the state outside of Baltimore;[9] fifth, low labor costs; sixth, low fuel costs; and seventh, access to and availability of capital for commercial startups. These conditions spurred the emergence of the poultry industry on the lower Maryland Shore."[10]

The changing landscape of commercial poultry drew the attention of the General Assembly of Maryland and ushered in a period of regulation. In 1937, it enacted legislation regarding poultry production and inspection in Maryland.[11] The General Assembly also created Boards and Advisory Councils to underscore the growing importance and economic benefit of poultry. In particular, the legislation empowered the State Board of Agriculture to "supervise the sale and transportation of poultry." The legislation further required anyone trading in poultry to procure a license from the State Board of Agriculture. In 1941, the General Assembly repealed and re-enacted the 1937 poultry legislation with amendments that further regulated commercial poultry production in Maryland.[12] The poultry legislations and amendments to legislations were aimed at commercial poultry producers although the early poultry pioneers had preferred deregulation.[13]

THE GREAT DEPRESSION PERIOD AND COMMERCIAL POULTRY PRODUCTION

The early phase of the commercialization of poultry came at a time when the agricultural sector of the economy was reeling from the effects of the Great Depression. The Great Depression of the 1930s impacted all sectors of the American economy including agriculture and commercial poultry. At the national level, the instability of the financial markets caused panic across the country as people lost confidence in the financial system. Although the drought had an impact on agriculture as a whole, it had a "lesser" negative impact on commercial poultry. Most of the crops that were by-products of agriculture required land, but commercial poultry was not tied as much to farmland as strawberries or corns and grains. Its ability to flourish was less restrictive as other crops.[14]

In its effort to aid the economy in its recovery from the Great Depression, the Federal government enacted the National Recovery Act (NRA) in 1933 with a Hatchery Code to regulate commercial egg hatchery production and ensure fair trading practices and price variances within the hatchery industry.[15] In 1935, the United States Supreme Court overturned the NRA, declaring it unconstitutional. Three years later, the Federal Trade Commission promulgated new rules and regulations. The new rules applied to the baby chick industry and prohibited "false, misleading, or deceptive methods" in intra-industrial competition.[16]

The commercial egg hatchery operations created opportunities for local growers to enter into contracts with commercial poultry companies. They raised the chicks for nine weeks by which time the chickens laid eggs and were consumable as meat.[17] The contractual obligations required the growers to provide the housing facility, which averaged more than $100,000 to construct.[18] In the 1930s and 1940s, the poultry houses on the Delmarva Peninsula were narrowly constructed and measured 15' to 21'w. The shed roof houses were about 150–400 feet in length and had no insulation.[19] In addition, the houses were constructed with fiberglass wall insulation and supplemented by exhaust fans to help maintain appropriate room temperature for the chickens.[20] Although these types of poultry houses were all over Delmarva, the same conditions were prevalent on the Lower Maryland Eastern Shore as there were no state lines that separated commercial poultry production on the Delmarva Peninsula. The housing specification and the attendant challenges such as cost and maintenance of the poultry houses contributed to fewer African Americans' involvement in commercial egg hatchery operations.

As commercial poultry gradually gained a foothold on the Lower Eastern Shore, Arthur Perdue became the dominant figure. He exemplified the financial ingenuity that was crucial during the early phase of the commercialization of poultry. He was financially frugal and did not cultivate a practice of borrowing money to invest or expand his business. Instead, he believed in building and expanding within his ability and rarely used credit. To this end, he only built new hatcheries, expanded, and purchased related items only when he had the money to do so. He purchased much of the chicken feed for his hatcheries in cash.[21]

Arthur Perdue's fiscal frugality is instructive. The early phase of commercial poultry emerged at a time of great financial difficulties. Families and businesses stretched the little money they had to meet their basic needs. Farmers or those in agriculture were in many instances the hardest hit during the depression because they had to contend with negative or unpredictable crop yields. Farmers had concerns over whether their crops would sell on the market. Money was an issue for many farmers. At the time, some farmers who borrowed money defaulted on their loans and many lost their farms. Perdue may have adopted a no-borrowing policy perhaps because he feared a possibility of defaulting on loans. Things were unpredictable, and it would have been too risky for him to rely on loans. But the uncertainty of early commercialized poultry seemed to have emboldened Perdue to find new ways to succeed. He had more to lose if he lost his business because of loan default.

Perdue's refusal to borrow money defied the standard business practice of capital acquisition for investment and expansion. He relied on the lessons that he learned from his father about saving, responsibility, and accountability to resist the urge to go into debt. In fact, Perdue recounted the time when he was 10 years old and his father gave him . . .

> 50 chickens to feed, care for and keep records on . . . [and] it was a way of teaching [me] to earn [my] own money and take care of it . . . I was able to make from $5 to $20 a month during the depression years when the average adult didn't make $5 a week.[22]

It seemed that the gift from Arthur Perdue's father was his first test at handling responsibility during times of economic adversity. It was a way of teaching Arthur about self-reliance and independence. It was a lesson well-learned that he took with him and used to his advantage. He literally changed the way people viewed and thought about poultry on the Lower Maryland Eastern Shore. One plausible reason that Perdue did not become debt-ridden was his recognition of the demand for poultry related products. Back in the 1930s, families kept small flocks in their backyards as a source of their daily or regular protein intake in the form of eggs. The flocks laid eggs for family breakfast and at the end of their egg laying years, which lasted for at least three to five years, they were either consumed or sold in the local open market. In this case, the typical chicken then served a double purpose. Perdue viewed them as long-term investments since there was always going to be a demand for poultry due to the public's need for protein rich food. The investment paid off because not only did families consume the eggs, but they also increased their consumption of chickens. In 1934, the per capita chicken consumption was 0.5 pounds.[23]

Despite the depression and the attendant economic devastation, commercial poultry production steadily evolved into an economic engine by the 1930s and 1940s. It helped ease the severity of the economic devastation on the Lower Maryland Eastern Shore. The Shore particularly benefited from the early phase of commercial poultry due to its proximity to "big city markets" like Philadelphia, New York City, and Baltimore. The large volumes of chickens that were transported to these "big city markets" helped the Lower Shore absolve the shock of the depression more than other places that relied more on agricultural crops like grains, strawberries, or tomatoes.[24]

To better appreciate and understand the context in which commercial poultry production emerged on Maryland's Lower Shore, a cursory examination of the political climate is imperative. Maryland's governor at the time of the depression, Albert C. Ritchie, a Democrat, was a man of profound pride and appeal. He was born in 1876 in Richmond, Virginia. He attended and graduated from the Johns Hopkins University in 1896. In 1915, he was elected Maryland's Attorney General. Four years later, Marylanders sent him to Annapolis by a 165-vote margin as their governor over Republican Harry W. Nice in one of the most closely contested elections in Maryland's history.[25] Prior to being elected governor, Ritchie had endeared himself to Marylanders as the man of the masses by going after public corruption when he served as attorney general and took on cases against energy giants such as the Gas and Electric Company of Baltimore, the forerunner of the Baltimore Gas and Electric.[26]

Governor Ritchie projected Maryland as a state with rugged individualism and pragmatism. He was supremely confident in Marylanders' ability to overcome the hardship caused by the depression. He believed strongly in local and state self-help relief efforts. For example, in 1931, an Unemployed and Relief Association was formed in the three Lower Shore counties to assist those who were negatively affected by the depression. The association, inspired by Governor Ritchie's insistence on self-help, met regularly, every fourth Thursday of the month where they held demonstration sessions for local farmers including poultry farmers to assist them with enhanced farming techniques.[27]

The agricultural demonstrations took place throughout the counties. In addition to helping local farmers, these demonstrations fostered a sense of collectivism among poultry farmers. There was also a Poultry Club, which met regularly at the Princess Anne Academy to share ideas about how to keep their chicks healthy and certified against infections.[28] For local farmers, it was imperative to keep their chickens healthy. When left unchecked, it was not unusual for bird infections to wipe out an entire flock of chickens. In the midst of the various self-help programs, there was still a need for federal assistance, but the governor insisted that Maryland was quite capable of handling the economic challenges.[29] But he finally gave in to the idea of outside assistance after much inter-governmental lobbying. Governor Ritchie attempted to use the assistance that Maryland received from the federal government by setting up a committee that would ensure that the relief monies were effectively distributed and that they actually reached those who were truly in need.[30]

Clearly, Governor Ritchie was a gifted public servant. However, he had two major shortcomings with regard to farmers. First, his pride may have clouded his views about the severity of the economic situation in the state in general and the Lower Maryland Eastern Shore in particular. In the midst of the economic difficulties that Marylanders faced, the governor spent time around the country canvassing for election to national office. His priorities did not portray him as an empathetic figure as many would have expected. His actions in initially turning down federal money further portrayed him as unconcerned, and his critics thought his ambitions had become antithetical to Maryland's interests.[31]

Second, Governor Ritchie's appointment of individuals who shared his views of economic reliance and recovery did not help poultry producers and farmers. Since the governor's appointees were beholden to his views, it would have been out of character for them to initiate programs that did not suit the governor's ideological position. In essence, the agricultural relief initiatives such as loans and grants meant to assist farmers including commercial poultry farmers did not reach the intended targets because relief aids in Maryland at the time were misdirected.[32]

When Franklin Delano Roosevelt became president in 1932, he brought new hopes of economic recovery and confidence to a fragile nation. His administration's relief programs called for a New Deal with Americans. The programs were geared towards the restoration of personal economic independence, societal and humanitarian programs, economic planning, and assertive leadership.[33] The New Deal programs under President Roosevelt, although well intentioned, did not solve the unemployment problem or end the depression.

In 1933, the U.S. Congress enacted the Federal Emergency Relief Act (FERA), which established the Federal Emergency Relief Administration and charged it with providing monies and other kinds of assistance to state and local relief efforts, many of which by this time had exhausted their relief funds. This agency portended better prospects for states like Maryland because they did not have to pay back the funds as they did loans. An important component of FERA was that states took on the responsibility for establishing the relief programs in their various states.[34]

Another relief initiative was the use of agricultural agents. On the Lower Maryland Shore counties, agricultural agents were sent in to assist the local farmers, and this proved helpful because they taught poultry farmers about improved methods of raising poultry. This initiative was particularly crucial since these lower counties were the epicenter of commercial poultry in Maryland. The objectives of these local agricultural agents were "to make the farmer contented, happy, and satisfied."[35]

Despite the economic difficulties that resulted from the depression, the Lower Maryland Eastern Shore was instrumental in Maryland's ability to produce 2 million commercial broilers in 1934.[36] The lessons and strategies that farmers had been taught by the agricultural agents were beneficial to the improvement of commercial poultry techniques. For example, there were two initiatives—the Egg Record Project and Grow Healthy Chicks Project. These strategies taught poultry farmers how to track the chicken eggs and ensure that their flocks were healthy. Both of these strategies assisted poultry farmers who helped raise the Lower Shore out of the depression.[37]

There were concerns about the sustainability of the emerging poultry economy of the Lower Maryland Eastern Shore during the depression.[38] The concern came about because poultry growers on the Lower Shore had not utilized the type of "rapid-growing strain of chickens" that ensured meatier quality and therefore improve economic viability. Researchers at the University of Maryland, College Park found that the progeny of a breeder was related to the growth of a particular strain of chickens. The researchers indicated that "long-shanked chickens grow more rapidly than short-shanked chickens."[39] With this finding, poultry growers were mindful to focus on raising long-shanked chickens, which helped to improve and enhance the viability of the industry in the region.

Through various programs, the federal government sought to assist Maryland with its relief efforts and spur the local and regional economic growth. However, mostly white farmers reaped the benefits of commercial poultry production because they owned most of the poultry farms on the Lower Maryland Eastern Shore.[40] The involvement of the federal government and the establishment of federal agencies during the depression were indicative of the extensive nature and impact of the depression and the local jurisdictions' inability to absolve the consequences. This realization brought about a shift in policies for American agriculture such as government sponsored farm subsidies, relief programs, and the exporting and marketing of key American agricultural products such as poultry.

For instance, in 1933, the federal government enacted the Agricultural Adjustment Act which provided subsidies for farmers and discouraged them from growing certain crops.[41] It also provided loans to farmers to produce certain staple crops with surplus to be released during periods of low yields to moderate prices. The subsidies for the farmers were generated through taxes imposed on cotton and tobacco.[42]

COMMERCIAL POULTRY IN THE POST-DEPRESSION ERA

The Great Depression occurred at a particularly significant time when many states in the country still struggled with the issue of racial identity and place in society. While the struggle for identity raged on the domestic front, America would soon be embroiled in a conflict overseas. Ironically, this conflict became an important factor that spurned the infant commercial poultry into a major business giant.

As the men and women of the armed forces fought in Europe during World War II, the U.S. government was faced with the challenge of feeding these troops and those of her allies in order to maintain high morale. The U.S. government attempted to boost morale by supplying poultry, more than 40 percent of which came from the Delmarva region. To this end, Delmarva poultry was in high demand by U.S. armed forces to the extent that the federal government, through the war-time agency, War Food Administration, took control of the production of poultry products until the end of the war.[43] Consequently, the demand for Delmarva poultry and, by extension, the Lower Maryland Eastern Shore was beneficial for commercial poultry production.

Nationally, in 1940, there was a significant increase in the gross income from overall commercial poultry, which accounted for about $939.9 million dollars. More than half of the total income, 62 percent came from egg marketing while 38 percent came from egg-laying and commercial chickens raised for consumption.[44] On the Lower Maryland Eastern Shore, in 1941, poultry producers registered early indications of a viable industry by earning a gross income of approximately seven million dollars.[45] In 1945, there was yet stronger evidence of an influential industry as poultry producers earned more than $2,930.4 million dollars in total gross income. Although the percentage of earnings that came from egg marketing decreased to 52 percent, the profit margin increased.[46]

In some ways, there was a regional specialization of sorts in commercial poultry production on the Lower Eastern Shore. Perdue built his first egg hatchery in 1930; however, the demand for poultry superseded his ability to meet the demand, much of which came from outside the Lower Shore. Thus, the New England region became a major supplier of eggs to the Lower Shore commercial poultry producers, and, in turn, the Shore seemed to have specialized in supplying dressed poultry, which brought big Northeastern restaurateurs to the Shore.[47]

One major distinction in this arrangement and why Lower Shore and Delmarva chickens were meatier than others was that the New England egg hatchers cross-bred their chickens. Cross-breeding, for example, New Hampshires and Plymouth Rocks, White Leghorns and Rhode Island Reds, created distinct types of chickens that were either used strictly for egg production or consumption.[48] Eastern Shore growers capitalized on these cross-breeding as well as in-breeding and specialized in producing meatier poultry meat. Consequent to these breeding, Lower Shore in particular and Delmarva poultry in general became famous for a distinct type of poultry meat that was in demand both at home and abroad.

As a result of the poultry exports to the battlefields of Europe and the Pacific, domestic consumption and demands were affected. Civilians at home were encouraged through the media to reduce their poultry consumption and give way to exports in an effort to boost the morale of the troops in the battlefield.[49] In fact, during the war period, eight out of every ten residents of the Delmarva region were involved in some aspect of the poultry industry, and, in just one year, the Delmarva supplied 18 million pounds of poultry to the U.S. armed forces.[50]

World War II impacted the black community on the Lower Shore, and, by extension, commercial poultry production because the community had been a major labor force for the industry. For example, as a result of the mandatory military draft, many blacks were drafted into the service and many were sent overseas. The draft deprived the black community of its many able-bodied men from these counties, where many had worked on poultry farms.

Despite blacks' involvement in commercial poultry production, some joined the armed forces out of frustration as many were confined to the labor underclass. Others joined for patriotic reasons, such as to defend the U.S. and fight the racism of Adolf Hitler. On the other hand, many were simply drafted. These black soldiers hoped that by fighting for democracy abroad they could extend democracy at home. Double V—victory abroad and victory at home—was a major goal of African Americans during World War II. The war also deprived many black families on the Lower Shore of the potential for upward mobility. In the men's absence, their wives and the women of the community stepped in and worked in the chicken factories, which by the 1940s had become a job magnet for the black community. Black women would later populate the rank and file of the frontline workers inside the poultry processing plants as a result of the absence of the black men.

The national importance and strategic relevance of poultry was most highlighted in the 1940s. The Eastern Poultry Institute distributed a motion picture documentary that traced the history of the Delmarva poultry industry and how it impacted World War II both at home and abroad.[51] In a sense, poultry served a domestic, military, and foreign policy objectives. Poultry served the domestic objective by giving consumers an alternative. Second, it served a military objective because poultry meat kept the soldiers from going hungry and perhaps deserting or becoming vulnerable to being captured by the enemy. Third, it served a foreign policy objective in the sense that European nations who received U.S. poultry exports were indebted to America's national goodwill. The exporting of poultry was good for America's image as a friend in need. After all, America was already a friend to Europeans as it was engaged in war which was said to be aimed at freeing them from tyrannical, political, social, and political oppression of Adolf Hitler, Joseph Stalin, and Benito Mussolini.[52]

Back on the domestic front, heavy and consistent economic activity spurred the local economies of the three counties in particular and the Delmarva region in general.[53] Although commercial poultry production increased, workers were generally paid lower wages.[54] For example, Maryland in general had increased poultry production to more than 31 million commercial broilers by 1947, with Somerset, Wicomico, and Worcester counties as important producers.[55] Also in the 1940s, Perdue switched from its emphases on layer chicks to broiler chicks in synch with the demand for consumable meat bird. This shift helped the company expand its profit margins as there was a high preference for poultry over red meat.[56]

Essentially, commercial poultry saved the Delmarva Peninsula along with the Lower Eastern Shore from near economic collapse. Its viability as well as the actions of the Federal Office of Price Administration proved crucial to recovery. To ensure that the armed services personnel had the food they needed, the federal government instituted a rationing policy of some food items including beef during the war era both to control price and to check supply. The policy led many Americans to switch to poultry consumption and gave the region a major boost in their post-depression recovery. Had it not been for government food purchases during World War II, perhaps the Lower Maryland Eastern Shore region would have taken longer to recover from the Depression. In order words, the battles in the European theatre benefited American commercial poultry in particular and agriculture in general. The money that U.S. government invested in poultry products assisted the Delmarva region in rebuilding its economic lifeline. The wartime economic purchases that brought jobs to many blacks were no longer taking place, but the post-war period also brought good economic tidings to the poultry industry as a whole as civilian consumers once again bought poultry in large quantities. There had previously been a government-imposed cap on civilian poultry sales and purchases during the war to ensure that American troops were getting sufficient poultry delivered on the battlefields.[57] Civilian poultry consumption meant continuation of jobs in the poultry industry.

To demonstrate the importance of poultry to the local economy, there were chicken house apartments during the 1940s, which housed chickens and workers in the same building but in different wings (see appendix). These apartments, prominent in Parsonsburg, Wicomico County, during World War II, served three purposes. First, they minimized the poultry farmers' cost of production by providing readily available housing for the poultry workers and reduced the commute time that it took if they lived a distance away from the farm. Second, these apartments gave the farm owners a sense of security in knowing that their chickens would be better cared for since the poultry workers lived in the same vicinity.[58] Third, the apartments minimized or eliminated the potential transmission of fungus or any other infectious diseases that may be brought to the chicken houses. Chickens are usually susceptible to acquiring diseases and then infecting the rest of the flock. Commercial poultry producers took every precaution to reduce the inevitable mortality rate.

The rural makeup of the Lower Maryland Eastern Shore counties made them more disposed to raising poultry and agriculture. Also, the participation of the then Maryland State College (MSC), which later became the University of Maryland Eastern Shore (UMES) and its agricultural experiment station helped local farmers acquire new techniques for farming. The residents also benefited from enhanced techniques and strategies developed by both Maryland State College's Maryland Cooperative Extension (MCE) and University of Delaware (UD) agricultural extension programs. The counties equally benefited from their proximity to the neighboring University of Delaware poultry science program, which had a poultry research and development program by the 1930s. In particular, Maryland State College's dual purpose of providing education for the region's black population, agricultural programs, as well as experiment stations for training farmers and conducting poultry research, were very instrumental in Somerset County's local economy.[59] The outcome of the research by MSC, MCE, and UD impacted the emergence of commercial poultry during the post-depression period. New techniques enhanced the quality and volume of production by both independent and contract commercial growers.

EMERGENCE OF POULTRY UNDERCLASS ON THE LOWER MARYLAND EASTERN SHORE

The poultry labor underclass on the Lower Maryland Eastern Shore emerged over time as commercial poultry production became a major player in the local economy. The underclass was in existence during the Great Depression. In the Works Progress Administration (WPA) programs in Maryland, for example, workers were categorized. Commercial poultry workers were classified as unskilled laborers and were required to perform 108 hours of work monthly in Somerset and Worcester counties and 122 hours a month in Wicomico. These laborers, most of them African Americans, were paid a monthly security rate of $32.40 in Somerset and Worcester counties and $36.60 in Wicomico County to work on the Eastern Shore Development Project, supervised by the Bureau of Agricultural Economics.[60] According to the 1930 Census of Agriculture, there were 178 poultry farms in Somerset, 290 in Worcester, and 186 in Wicomico Counties.[61]

In the 1930s and 1940s, African American involvement in the commercial poultry industry was largely confined to cumbersome and energy-sapping activities. Their involvement in commercial poultry production was mostly through their work in the processing plants where they slaughtered, cleaned the chickens, and readied them for shipment to designated destinations.[62] In 1945, non-whites accounted for 11.2% of farm operators compared to 88.8% of white poultry farm operators (as shown in Table 3.1).[63] There were few black poultry farm owners. In some instances, blacks managed the poultry farms for whites and were paid for their services as farm managers.

By 1945, 1,645 farms in Somerset, Wicomico, and Worcester counties reported a total poultry value of $24,791,536 million with the most farms and value in Worcester County. According to the 1940 census, blacks were 31.4% of Worcester County population. It can be surmised that the farms that reported their net value in 1945 relied on the sizable black population for slaughtering, cleaning, and readying the chickens for shipment.

One factor in the emergence of the poultry labor underclass was a lack of education. For example, the 1930 and 1940 educational censuses showed that the majority of workers in the poultry industry were either high school graduates or drop outs. Many workers were driven into the underclass because of educational limitations. For African Americans, poorly funded schools and a high rate of illiteracy were important factors that contributed to their role as members of the labor underclass. Many young blacks left school and went into the world of work where most of the works available to them were in farming and manufacturing. Most of them worked in these jobs because they were unskilled. On their part, whites mostly held the skilled jobs because they attended better funded schools and therefore were more prepared. In addition, whites received training in other areas besides agriculture and farming.

Table 3.1. Poultry Farm Operators in Maryland, 1945

	State	*Percentage*
All poultry farm operators	102,917	100%
Total White	91,358	88.8%
Total non-white	11,559	11.2%

Source: U.S. Census of Agriculture, 1945, Special Report; Maryland, Table 13, 173.

Table 3.2. Number and Value of Poultry Farms in the Three Lower Eastern Shore Counties, 1945

	State	*Somerset*	*Wicomico*	*Worcester*
Farms reporting	4,972	331	648	666
Total value ($)*	37,346,057	3,750,351	8,749,583	12,291,602

*in millions

Source: U.S. Census of Agriculture, 1945, Maryland; County Table VIII, 93 8.

Minimal education led to minimal wages and the low wages created a state of dependency. The companies paid their workers low salaries in exchange for job security. This arrangement had two benefits. One, it ensured the continued flow of profits for the companies by paying their workers less and gaining huge profits because there was a demand for poultry. Second, workers were reasonably assured of employment because consumers always needed nutritional balance in their diet, which meant that they sought out beef, poultry, pork, or any other type of meat to meet this nutritional requirement. Poultry work, therefore, gave the workers a sense of job security that was not available with other jobs.

The workers were forced to stop attending school for various reasons, chief among which was the financial situation of their families. Students from the age of 7 years through 13 years started off attending school. As they grew older and the pressure to work and bring in money to support their families persisted, many left school to work. As for blacks, they had a higher rate of illiteracy than whites, and, without an education, a labor underclass dominated by blacks was inevitable.

Clearly, blacks on the Lower Maryland Eastern Shore counties did not fare well in education in the 1930s when compared to whites. The illiteracy rate for blacks at 23.7% was highest in Worcester County. In Somerset County, for example, 16.6% of the black population, 10 years and above, about half of the black population, were classified as illiterate (10 years and above). In Wicomico County, 12.9% were illiterate. Even for literate blacks who had a high

Table 3.3. Illiteracy Rates of the Lower Eastern Shore Counties for People Aged 10 Years and Over (1930)

Race	*State*	*Somerset*	*Wicomico*	*Worcester*
Total	49,910	18,694	25,487	17,294
	3.8%	7.6%	5.8%	10.4%
Whites	11,561	383	788	611
	1.4%	3.1%	3.9%	5.0%
Blacks	25,073	1,032	673	1,189
	11.4%	16.6%	12.9%	23.7%

Source: U.S. Census, 1930; Maryland, Table 13, 1056.

Table 3.4. Attendance and Completion Rates of the Lower Eastern Shore Counties for People Aged 7 to 25 Years Old (1940)*

	Somerset	*Wicomico*	*Worcester*
Attending School			
7–13 years old (total)	2,275	4,009	2,712
(N) attending	2,648	3,746	2,592
%	96.8%	93.4%	95.6%
14 and 15 years old	853	1,217	819
(N) attending	666	980	625
%	78.1%	80.5%	76.3%
16 and 17 years old (total)	864	1,214	791
(N) attending	414	616	326
%	47.9%	50.7%	41.2%
18–20 years old (total)	1,154	1,948	1,141
(N) attending	145	248	108
%	12.6%	12.7%	9.5%
21–24 years old (total)	1,253	2,418	1,393
(N) attending	20	47	20
%	1.6%	1.9%	1.4%
Years of School Completed			
Male, 25 years and over	5,957	10,138	6,105
No school years completed	275	392	384
Female, 25 years and over	5,840	10,232	5,949
No school years completed	196	273	229

N = Not attending

*In the 1940 Census, the information that was gathered focused on school attendance and completion. The information captured the number of students attending and completing school by age and county.

Source: U.S. Census of Population, 1940; Maryland, Table 21.

school education, they had limited employment opportunities because many companies only hired blacks for menial jobs.[64]

These statistics are instructive because they indicate a preponderance of blacks in the poultry labor underclass. As well, their illiteracy rate stunted their upward mobility. The high illiteracy rates in the counties in proportion to black populations helped to create a poultry labor underclass during the early phase of commercialized poultry production on the Lower Maryland Eastern. Lower Shore blacks became the poultry labor underclass because their high illiteracy rates limited their occupational choices.

In 1940, the U.S. Bureau of the Census modified the variables for measuring education. The data only captured the number of students attending and completing school. Across the races, there was widespread fluctuation in school attendance and completion, perhaps due to the post-Great Depression economic reconstruction. Wicomico County led the other two counties with the number of children attending and completing school. The county led in this category mainly because it was more urbanized than Somerset and Worcester Counties. In the same year, there were more males 25 years and over, who did not complete school in Somerset County than females in the same age bracket. Although completing school did not secure them better-paying jobs, those that completed school at least improved their chances of competing for other employment opportunities.

In the three Lower Eastern Shore counties, most of the poultry farmers and farm owners were whites. However, "there were a small number of black poultry owners who inherited their farms from previous generation of black farm owners."[65] These generally were subsistent farms. According to the U.S. Census of Agriculture, in 1930, only twenty poultry farmers owned less than 3 acres of land in Somerset County on which to grow their chicken feed while there were only three farmers in Wicomico and twelve in Worcester counties (information on ownership by acreage by race is not available).

This table of ownership by acreage suggests that there were clear and distinct social and economic classes within the counties. Forty-three farmers owned between 50 and 99 acres of land in Somerset County, while there were fifty-four farmers who owned the same acreage in Wicomico County and in Worcester County there were fifty-seven farmers with the same acreage. In essence, most poultry farmers in these counties in 1930 owned between 20 and 99 acres of poultry farm.

Ownership of poultry farms on the Lower Maryland Eastern Shore translated to economic power. In the instance where blacks owned land, it was typically less than the average acreage. For blacks, owning the same amount of poultry farm acreage as whites meant economic power, parity, and independence. Black ownership of poultry farm was not common.[66] This region mimicked the national pattern of limited black economic independence.[67]

Table 3.5. Acreage of Poultry Farms on the Lower Eastern Shore Counties, 1930

Type and size	*State*	*Somerset*	*Wicomico*	*Worcester*
Under 3 acres	173	20	3	12
3-9 acres	480	26	29	29
10–19 acres	391	15	14	37
20–49 acres	830	35	48	103
50–99 acres	597	43	54	57
100–199 acres	356	24	33	43
175–259 acres	75	8	4	5
260–499 acres	33	4	1	4
500–999 acres	5	3	—	—
1,000–4,999 acres	1	—	—	—

Note: The same variables for poultry farm acreage in the 1930 census were unavailable in the 1940 Census.

Source: U.S. Census of Agriculture, 1930; Maryland, Volume III, Part 2; County Table V, 95.

There were instances where blacks were bequeathed large acres of lands but lost them to eminent domain and confiscated for public use. For example, in Berlin, [Worcester County], some public works projects like roads were built through black-owned land appropriated under the right of eminent domain.[68] Due to the significance of land, some of the black residents interviewed in this study were of the belief that local officials invoked eminent domain as a way of minimizing blacks' chances of land ownership.

In contrast to Somerset and Wicomico counties, Worcester County was more rural and had fewer urban attractions at that time, hence making it an ideal place to practice commercial poultry. Also in 1935, 2,271 poultry farms reported raising 217,031 chickens over 3 months old in Wicomico, explaining perhaps the beginning of the dominance of the county in commercial poultry production as well as the labor pool that was required to produce at high capacity. In this sense, African Americans, notwithstanding their population when compared to Somerset and Worcester Counties, still comprised majority of the labor underclass in commercial poultry production.

Between 1930 and 1935, Wicomico and Worcester Counties reported the most poultry farms and chickens three months or older raised. Wicomico figured highly in the farms reporting category in part because of its growing urbanization and its ability to attract more potential and viable poultry growers who would qualify as contract growers for the bigger companies.

Before the early phase of the commercialization of poultry on the Lower Eastern Shore, there had been a pattern of black domination of the labor underclass when they worked fruit and vegetable industries. Many African

Table 3.6. Total Rural Farm Population of the Lower Maryland Eastern Shore Counties, 1930

	Total	*Whites*	*Blacks*
Somerset	7,433	4,741	2,692
Wicomico	9,924	8,105	1,819
Worcester	10,402	7,331	3,071

Source: U.S. Census of Population, 1930; Maryland.

Table 3.7. Total Rural Farm Population of the Lower Maryland Eastern Shore Counties, 1940

	Total	*Whites*	*Blacks*
Somerset	6,341	4,096	2,199
Wicomico	10,148	7,738	2,294
Worcester	8,239	5,880	2,291

Source: U.S. Census of Population, Vol. II, part 3, Maryland, 1940.

Table 3.8. Poultry Farms and Production on the Lower Maryland Eastern Shore, 1930–1935

	Year	State	Somerset	Wicomico	Worcester
Chickens over 3 months old	1930	3,777,072	175,482	224,191	267,632
	1935	3,589,071	124,988	217,031	207,289
Farms reporting	1930	39,648	1,369	1,946	2,043
	1935	39,416	1,205	2,271	1,790
Chickens raised	1934	7,030,654	371,942	404,303	626,949
Farms reporting	1934	36,059	1,125	2,126	1,671

Source: U.S. Census of Agriculture, 1935, Maryland, County Table V, 420-1.

Americans worked in the tomato canneries or picked tomatoes for white farmers and farm owners. Ordinarily, tomato pickers were paid twenty-five cents for a basket of tomatoes.[69] However, black pickers, mostly women, were paid six cents per basket in the 1930s. After the tomatoes were gathered, they were transported to the cannery where another shift of black pickers peeled and readied them for the next phase. The tomatoes were boxed or sent to the assembly line to be grounded and canned before being shipped to other areas of the region. Many workers left the canneries and went to work another shift in the poultry processing plants for long hours to earn enough money to survive.

Over time, these tomato pickers became the permanent underclass as line workers in the commercial poultry plants, which operated year round compared to tomato production, which was seasonal.[70] Black women did most of the tomato picking and processing while black men worked in the more labor intensive lumber industry. As the commercial poultry industry gained momentum, it drew African Americans to its labor pool. Despite their underclass status, the black community on the Lower Eastern Shore had pride, flexibility, and a willingness to take on jobs of different sorts to keep families and communities intact. Many blacks did poultry work on the lower Shore in the 1930s and 1940s because it was available. Generations of early black poultry workers in Salisbury [Wicomico] and Showell and Berlin [Worcester] where the factories were located contributed to the growth of the industry as well as to the regional and state economy.[71] In essence, there was a pattern of blacks being the labor underclass on the Lower Maryland Eastern Shore well before commercialization of poultry eclipsed other sectors of the local economy.

To be sure, life for blacks in the three Lower Shore counties was harsh. The majority of them lived in the rural areas where affordable and standard housing were major issues of concern. Similar hardship about affordable housing also befell blacks across the country. They organized to try to improve their housing conditions. An organization called the Southern Tenant Farmers

Union (STFU), which had mixed racial membership, was founded in 1934 as a voice for the tenant farmers in the South, many of whom were black.[72] Henry L. Mitchell and Henry Clay East of Arkansas founded the STFU and sought to break the near monopoly of white farm ownership.[73] Many of these tenant farmers were victimized by whites who misappropriated monies that were intended for their assistance. The STFU agitated for, among other things, decent living arrangements for farm laborers and tenants. The formation of this union threatened white planters who opposed the union because it was one of the few instances when blacks and whites joined together to confront a common issue.[74]

The STFU did not gain a foothold on the Lower Maryland Shore for a number of reasons. There was not an organized and cohesive black agricultural pressure group and during this time, most blacks in these counties were more preoccupied with survival than making a political or social statement. In addition, there were minimal transportation links with other parts of the state and beyond that would have made it easier for union officials to travel from the Deep South to Maryland to organize black workers.

Instead of relying on outside organizations, blacks on the Lower Shore counties looked and worked from within to overcome difficulties in housing. Communities rallied together and individuals with bigger homes and more resources shared with those who had little. Although blacks were not wealthy in comparison to whites, some families had bigger homes and simply shared with those in need. In addition, two key factors, faith and family, kept these families together and helped them overcome their struggles with the economy. Holden stated:

> Throughout the Great Depression years—when banks were folding all over the United States; when millions of families were being wiped out financially; when newspapers were headlining suicides of the bereft; when breadlines were lengthening in big cities; when men were working fifty cents to a dollar a day—the worsening state of black poverty was simply taken for granted. Neither recovery programs nor the "We do our part" slogan addressed the growing needs of black people, many of whom already had done their part without even token recompense. Therefore, blacks were left to carve out a means of survival—one that demanded extraordinary determination, stamina, ability, and dreams of a better life—all fortified by the same religious faith that sustained our ancestors. It is that priceless part of our heritage that ties us to the faith of our fathers, to the church, and to the belief that with God on our side, we shall overcome.[75]

Thus, stable black families were essential to the early phase of the commercialization of poultry production. The stability of black families even in the face of a devastating economic depression was on display in the 1930s and 1940s as they shaped the commercialization of poultry on the Lower Eastern

Shore with their involvement as the labor poultry underclass. Not only did they overcome the depression, they helped build the early phase of the commercial poultry industry in Maryland.

CONCLUSION

During the early phase of the commercialization of poultry in the Lower Eastern Shore counties, there were three identifiable conditions and problems that affected poultry farming. First, there was insufficient food and feed for farmers and livestock due to poor management; second, there were bank failures and general depression; and third, there was a lack of foresight and county pride in the effectively showcasing the broader significance of commercial poultry in these counties. Poultry farmers had to alter their techniques to take advantage of their new markets of poultry consumption.

To remedy these conditions and problems, the agricultural agent at the time, Louis H. Martin recommended several programs including live at home program; employment with low wages; a five-step program to be sponsored by the Maryland Extension Service, and education and training for everyone including non-farmers through community, county, and state meetings as solutions to the poultry conditions on the Lower Shore counties.[76]

The mass education and training took place because a large number of these county residents were involved in raising poultry even at very young ages. For example, elementary school students were taught about poultry husbandry and then what they learned in school was reinforced at home in the poultry houses. It seemed that the children had ambitions of someday engaging in commercial poultry. They wrote letters in which they expressed their pride in and satisfaction with the poultry education and demonstrations conducted by agent Martin. The training sessions helped them improve the health and condition of their chickens.

The periods of the 1930s through the 1940s were the building phases of commercial poultry production in Maryland. Although blacks did not own most of the poultry farms, their labor was essential to the process of commercialization of poultry production. Despite the widespread negative economic impact brought on by the Great Depression, the early phase of commercial poultry flourished on the Lower Maryland Eastern Shore largely because producers exploited a consumer demand for poultry meat. The prominence of poultry during the depression aided the regional economic recovery. The Lower Maryland Eastern Shore figured more prominently in the early phase of commercial poultry because it was part of the birthplace of commercial poultry on the East Coast United States: the Delmarva Peninsula.

Also, World War II was economically beneficial to the early phase of commercial poultry production on the Lower Maryland Eastern Shore. The U.S. armed forces' demand for Delmarva chickens injected much needed economic activity in an area that was severely impacted by the great depression. However, for both Delmarva and the Lower Maryland Shore to meet this high demand, it relied on the involvement of the labor underclass, most of whom were black, to meet the supply. The Delmarva region as a whole helped the Lower Shore emerge as the epicenter of Maryland poultry.

The early phase of the commercialization of poultry was a strategic period which laid the foundation for the industry. It was a trial and error period that allowed producers to develop new ways to produce and market poultry while presenting it as a preferred alternative to red meat, which had more fat content and therefore placed consumers at a higher risk for coronary-related diseases. The early phase further positioned commercial poultry into the next phase of development and consolidation into a much larger-scale commercialization where vertical integration transformed the industry.

NOTES

1. Walter H. Gardiner and John W. Wysong, "Combining Farm and NonFarm Labor Employment Opportunities," *Agricultural and Resource Economics Information Series*, [No. A-1], 1–2; Box 8, Contribution Nos., 5000–5249, Maryland Agricultural Experiment Station, October 1975 (Special Collections, University of Maryland Libraries, College Park).

2. Ibid.

3. See T. B. Symons, "Maryland Poultrymen and National Conservation Program," in Papers of Morley A. Jull [Articles Manuscripts, 1947] Series II, Box 1, Folder 3 (Special Collections: University of Maryland Libraries, College Park, Maryland).

4. See the following works, John Steinbeck, *The Grapes of Wrath*, First Edition (New York: The Viking Press, 1939); Paul Bonnfield, *The Dust Bowl: Men, Dirt, and Depression* (Albuquerque, New Mexico: University of New Mexico, 1978); John Opie, *The Law of the Land: Two Hundred Years of American Farmland Policy* (Omaha: University of Nebraska Press, 1994); Katelan Janke, *Survival in the Storm: The Dust Bowl Diary of Grace Edwards, Dalhart, Texas, 1935*, *Scholastic* (September 2002), Siegfried D. Schubert, Max J. Suarez, Philip J. Pegion, Randal D. Koster, and Julio T. Bacmeister, "On the Cause of the 1930s Dust Bowl," *SCIENCE* Magazine, (19 March 2004), 1855–1859 and Timothy Egan, *The Worst Hard Time: The Untold Story of Those Who Survived The Great American Dust Bowl*, (New York: Houghton Miflin Company, 2006).

5. George Watts and Connor Kennett, "The Broiler Industry," *Poultry Tribune* (1995), 8.

6. Frank Gordy, *A Solid Foundation . . . The Life and Times of Arthur W. Perdue* (Salisbury, Maryland: Perdue Incorporated, 1976), 24.

7. Ibid, 32.

8. William H. Williams, *Delmarva Chicken Industry: 75 Years of Progress* (Georgetown, Delaware: Delmarva Poultry Industry, 1998), 9.

9. "Community Economic Inventory: Wicomico County, Maryland." (Annapolis, Maryland: Division of Business and Industrial Development, Maryland Department of Economic and Community Development, 1974), 30–31.

10. *Maryland's Historic Somerset*. (Princess Anne, Maryland: Board of Education of Somerset County, 1969), 28. The conditions that were prevalent in Somerset County were also existent in the other two counties—Wicomico and Worcester. See also William H. Williams', *Delmarva's Chicken Industry . . .* esp. Chapter 2.

11. *The Annotated Code of the Public General Laws of Maryland*, 1930; Volume 379, Chapter 347, Sections 147–152 [Maryland State Archives].

12. [Senate Bill 353] Maryland General Assembly, *Annotated Code of Maryland*, 1941; Article 48, Section 168.

13. Edward H. Covell, "The Broiler Industry—Then, Now and Tomorrow," Unpublished paper (Archives of Delmarva Poultry Industry, Inc., Georgetown, Delaware).

14. "The Broiler Industry," *A Report by the National Broiler Council*; Research 1985. The word "lesser" is used here in a limited sense to mean that the crops that are produced on land suffered more devastating consequences in terms of the farmers' survivability in the event of extreme or extended natural disasters like famine and drought that will force them out of business. These natural disasters had more severe consequences for full-time farmers than a 25 percent mortality rate for commercial poultry farmers who still could salvage the remaining 75 percent of their flock.

15. John L. Skinner, (ed.), *American Poultry History, 1823–1973* (Madison, Wisconsin: American Printing and Publishing, Inc., 1974), 174.

16. Ibid.

17. Frank Perdue, interview in *Fortune Magazine*. Available at http://www.fortune.com.

18. Jerry Truitt, interview by author, Georgetown, Delaware. Mr. Truitt was a former Director of the Delmarva Poultry Industry Inc., and also a retired banker. He was directly involved in commercial lending to Delmarva poultry growers during the large-scale poultry commercialization period in the 1950s. See also, Donald D. Stull and Michael J. Broadway, *Slaughterhouse Blues: The Meat and Poultry Industry in North America* (Belmont, CA: Thompson/Wadsworth, 2004), 46.

19. George W. Chaloupka, "The Early Days of Our Local Poultry Industry," Unpublished paper from his Private Collections (paper is in the author's possession).

20. William L. Henson, "The U.S. Broiler Industry: Past, Present Status, Practices, and Cost," *Agricultural Experiment and Research Services*, USDA, No. 149 (May 1980), 30.

21. Gordy, 38–40.

22. Ibid, 35.

23. Verel W. Benson and Thomas J. Witzig., "The Chicken Broiler Industry: Structure, Practices, and Costs," *Agricultural Economic Report No. 381*, (Washing-

ton, D.C.: United States Department of Agriculture, Economic Research Service, 1977), ii-1.

24. By "big City markets," I am referring to more urbanized areas with populations of more than 50,000 people. Because of their size, it can be surmised that these cities needed a high and steady supply of poultry to meet the urban demand.

25. Robert J. Brugger, *Maryland: A Middle Temperament, 1634–1980* (Baltimore, Maryland: The Johns Hopkins University Press, 1988), 451.

26. James B. Levin, "Albert C. Ritchie: A Political Biography." (Ph.D. dissertation, City University of New York, 1970), 25–39.

27. General Records of the Department of Agriculture, Record Group 33, Extension Service Annual Reports, T-865, microfilm roll number 36, National Archives Building, College Park.

28. Ibid.

29. Charles M. Kimberly, "The Depression in Maryland: The Failure of Voluntaryism," *Maryland Historical Magazine*, Vol. 70, No. 2 (Summer 1975), 189–202.

30. Robert J. Brugger, *Maryland: A Middle Temperament, 1634–1980.* (Baltimore: The Johns Hopkins University Press, 1988), 490–551.

31. Ibid.

32. For an in-depth analysis of the economic conditions in the State during the Great Depression, see Charles M. Kimberly's full dissertation titled, "The Depression and the New Deal in Maryland" (Ph.D. diss., American University, 1974).

33. See the following, Arthur Meier Schlesinger, *The Coming of the New Deal* (New York: Houghton Mifflin, 2003); Patricia Sullivan, *Days of Hope: Race and Democracy in the New Deal Era* (Chapel Hill: University of North Carolina Press, 1996); Roger Biles, *The South and the New Deal* (Lexington: University Press of Kentucky, 1994); Herbert Hoover, *American Ideals verses the New Deal* (New York: Scribner Press, 1936), and Suzanne Ellery Chapelle and Glenn O. Phillips, *African American Leaders: A Portrait Gallery* (Baltimore: Maryland Historical Society, 2003), 32, among other works.

34. John Mack Faragher, Mari Jo Buhle, Daniel Czitrom, and Susan H. Armitage, *Out of Many: A History of the American People*, Second Edition (Upper Saddle River, New Jersey: Prentice Hall Inc., 1997), 761–2. See also the following works, Robert S. McElvaine, (ed.), *Down and Out in the Great Depression: Letters from the Forgotten Man* (Chapel Hill: University of North Carolina Press, 1983); Edwina Amenta, *Bold Relief: Institutional Politics and the Origins of Modern American Social Policy* (Princeton: Princeton University Press, 2000); Raymond Wolters, *Negroes and the Great Depression: The Problem of Economic Recovery* (Westport, CT: Greenwood Publisher Group Inc., 1974); David M. Kennedy, *Freedom from Fear: The American People in Depression and War, 1929–1945* (New York: Oxford University Press, 1999); John Joseph Wallis, "Employment, Politics, and Economic Recovery During the Great Depression," *The Review of Economics and Statistics*, Volume 59 (August 1987), 516–520; D. C. Reading, "New Deal Activity and the States," *Journal of Economic History*, Volume 36 (December 1973), 792–810; Donald Howard, *The WPA and Federal Relief Policy* (New York: Russell Sage Foundation, 1943), Jim F. Couch and William F. Shuggart, "New Deal Spending and the States: The Politics of

Public Works," In J. Heckelman, J. Moorhouse, and R. Whaples, (eds.), *Public Choice Interpretations of American Economic History* (Boston: Kluwer Academic Publishers, 2000), 105–122, and John Wallis, Price Fishback, and Shawn Cantor, "Politics, Relief, and Reform: The Transformation of America's Social Welfare System during the New Deal," (unpublished paper) among other works.

35. Extension Service Annual Report [Somerset, Wicomico, and Worcester Counties, 1933] Record Group 33, National Archives Building, College Park, Maryland.

36. John P. Trimmer, *Agricultural Maryland: A Sketch of Free State Farming* (College Park: Maryland Agricultural Extension Service, 1949).

37. Record Group 33, [Narrative Report of C. Z. Keller, County agent for Somerset County, 1932, p. 31], National Archives, College Park, Maryland.

38. See Morley A. Jull, "Will Your Birds Be Efficient Enough If A Depression Comes?" Papers of Morley A. Jull, [Article Manuscripts, 1945], Series II, Box 1, folder 3 (Special Collections: University of Maryland Libraries, College Park, Maryland).

39. Ibid.

40. See C.W. Pierce, "An Economic Study of 99 Maryland Poultry Farms," (M.S. Thesis, University of Maryland, College Park, 1933), 3.

41. For a more in-depth discussion on this and related agencies, see the following works, Michael D. Bordo, Claudia Goldin, and Eugene N. White, (eds.), *Defining Moment: The Great Depression and the American Economy in the Twentieth Century* (Chicago: University of Chicago Press, 1998); Congress, House of Representatives, Committee on Agriculture, *Hearings before the Committee on Agriculture,* 84th Congress, 1st Session and passim; Sidney Baldwin, *Poverty and Politics: The Rise and Decline of the Farm Security Administration* (Chapel Hill, North Carolina: University of North Carolina, 1968), and James T. Young, "The Origins of New Deal Agricultural Policy: Interest Group's Role in Policy Formation," *Policy Studies Journal,* Volume 21, Issue 2 (June 1993), 190–209.

42. See the following, Edwin G. Nourse, Joseph S. Davis, and John D. Black, *Three Years of the Agricultural Adjustment Administration* (Washington, D.C.: Brookings Institution, 1937); Alan Brinkley, *The End of Reform: New Deal Liberalism in Recession and War* (New York: Vintage Books, 1996; Alan Brinkley, *American History: A Survey,* 10th Edition (New York: McGraw Hill College, 1999), and Congress, Senate, *Agricultural Adjustment Act of 1937*, 75th Congress, 1st sess., (May 18, 1937). The U.S. Supreme Court ruled in 1936 that the Act was an unconstitutional attempt by the executive branch of the federal government to take over the duties of the legislative branch. It opined that the federal government could not regulate agriculture through the act and that taxes could not be levied on one group for the benefit of another. The Court, however, accepted the second AAA enacted in 1938, after the provisions in the new legislation stipulated that the revenue for the subsidies be generated through general taxation.

43. John Steele Gordon, "The Chicken Story" *American Heritage* (September 1996), 66.

44. John Stanley Stiles, Jr., "Comparative Costs of Cutting and Packaging Poultry," (M.S. Thesis, University of Maryland, College Park, 1959), 1.

45. Thomas J. Davies, "The Broiler Industry in Maryland," (M.S. Thesis, University of Maryland, College Park, 1942), 8.

46. Bureau of Agricultural Economics, *Farm Production, Disposition and Income from Chickens and Eggs, 1945–1949*, [United States Department of Agriculture, BAE 268], (Washington D.C.: U.S. Government Printing Office, 1952), especially pages 11, 14, and 16.

47. William H. William, *Delmarva Chicken Industry. . .* , 28–33.

48. See the following works, "Practical Breeding Methods to Produce Superior Quality Chicks," Lecture presented by Morley A Jull (29 August 1946); "The Problem of Inbreeding," "How the Research Program at the University of Maryland Aids Maryland Poultrymen," "Chicken of Tomorrow—Eastern Viewpoint," New Problems facing Maryland's Poultry Industry," in Papers of Morley A. Jull, [Article Manuscripts 1946], Series II, Box 1, Folder 2, (Special Collections: University of Maryland Libraries, College Park, Maryland).

49. Ben Kerner, *Your Chicken Has Been To War* [a motion picture documentary]. Dover: Delaware State Archives, Division of Historical and Cultural Affairs, Hall of Records, 1943.

50. Ibid.

51. Ben Kerner, *Your Chicken Has Been To War. . . .*

52. Ibid.

53. Kimberly R. Sebold, "The Delmarva Broiler Industry and World War II: A Case in Wartime Economy," *Delaware History* Vol. XXV, No. 3 (Spring-Summer 1993): 200–214. See also, Morley A. Jull, "Broiler Prospects in Early 1945," Articles Manuscripts by Morley A. Jull, 1945, Series II, Box 1, Folder 1 (Special Collections: University of Maryland Libraries, College Park, Maryland).

54. By the 1940s, poultry processing plant workers earned less than 60 cents an hour. While the Fair Labor Standards Act of 1938 mandated a graduated wage increase to 60 cents an hour within two years of passage of this Act, the profits from the booming poultry industry did not trickle down to the workers.

55. John P. Trimmer, *Agricultural Maryland* . . . 22–3.

56. S. Lesson and J.D. Summers, *Commercial Poultry Nutrition* (Guelph, Ontario: University Books, 1991), 2–3.

57. William H. Williams, *Delmarva Chicken Industry* . . . 35–45.

58. Kimberly R. Sebold, "Chicken-House Apartments on the Delmarva Peninsula," *Delaware History*, Vol. XXV, No. 4 (Fall-Winter 1993–4).

59. John R. and Ruth Ellen Wennersten, "Separate and Unequal: The Evolution of a Black Land Grant College in Maryland, 1890–1930," *Maryland Historical Magazine*, Volume 72, No. 1 (Spring 1977), 110–117.

60. Series I: WPA Administrative Records, Maryland WPA, Schedule of Hourly Wage Rates, Hours of Work, and Monthly Earnings [Microfilm] (Special Collections: University of Maryland Libraries, College Park).

61. C.W. Pierce, "An Economic Study of 99 Maryland Poultry Farms," (M.S. Thesis, University of Maryland, College Park, 1933).

62. There was a wage differential in the processing industry just as in other manufacturing industries. White poultry workers benefited from this wage differential. For example, Records of the National Recovery Administration (RG 9, classified general files of the NRA) contained "Wage Differentials—Negro and White" indicating that whites earned more than blacks.

63. In the 1940 census enumeration, non-whites included Chinese, Japanese, Indians, and African Americans. Most "non-white" Marylanders were African Americans.

64. Spotwood Jackson, interview by author, Baltimore-Salisbury, Maryland.

65. Anna Smith, interview by author.

66. Ibid.

67. See Elizabeth Wright, "Without Commerce and Industry, the People Perish," *Issues and Views*, (Spring 1991); Available at http://www.issues-views.com, and C. Mason Weaver, "Too Much Political Power, Not Enough Economic Independence," [Commentary], Washington D.C.: *National Center for Public Policy Research*, (January 1997).

68. Eminent domain is the power of the government to take someone's property and convert it to public's benefit. Usually, the government offers reasonable compensation for the seizure. In some instances, blacks did not receive compensation. Anna Smith, interview by author. Mrs. Smith's claim was corroborated by other residents.

69. Adele V. Holden, interview by author.

70. Ibid. For more discussion on fruits and vegetable pickers in the Southeast United States, see, Charles D. Thompson and Melinda F. Wiggins, (eds.), *The Human Cost of Food* (Austin: University of Texas Press, 2002); Phillip L. Martin, *Promise Unfulfilled: Unions, Immigration, and the Farm Workers* (Ithaca: Cornell University Press, 2003), and Congress, House of Representatives, Committee on Agriculture, *Farm Labor: Hearing before the Committee on Agriculture,* 80th Congress, 2nd Session, H.R. 6819 and S. 2767, (11 June 1948).

71. Adele V. Holden interview by the author.

72. Mark Fannin, *Labor's Promised Land: Radical Visions of Gender, Race, and Religion in the South* (Knoxville: The University of Tennessee Press, 2003): 71–129. Although the leadership did not wholeheartedly embrace black equality, the STFU was still viewed more positively by black farmers and race ultimately became an issue of contention. See also 221–253.

73. Ibid, 71–129. See also James Gilbert Cassedy, "African Americans and the American Labor Movement," *Prologue: Quarterly of the National Archives and Records Administration* Volume 29, No. 2 (Summer 1997); and The Eleanor Roosevelt Paper. "Southern Tenant Farmer's Union," in *Teaching Eleanor Roosevelt*, ed. Allida Black, June Hopkins et al. (Hyde Park, New York: Eleanor Roosevelt National Historic Site, 2003).

74. John Hope Franklin and Alfred A. Moss Jr., *From Slavery to Slavery: A History of African Americans* (New York: McGraw-Hill Companies, Inc., 2000), 433–439.

75. Adele V. Holden, *Down on the Shore: A Memoir, The Family and Place that forged a Poet's Voice* (Baltimore: Woodholme House Publishers, 2000), 85. See also the important work about black families in the South including Lower Shore counties by William W. Falk, *Rooted in Place: Family and Belonging in a Southern Black Community* (Piscataway, New Jersey: Rutgers University Press, 2004).

76. See Extension Service Annual Report [Somerset, Wicomico, and Worcester Counties, 1933] Record Group 33, National Archives Building, College Park, Maryland.

Chapter Four

Development and Consolidation of Large-Scale Commercial Poultry Production, 1950s to 1990s

Commercial poultry production on the Lower Maryland Eastern Shore went through a major transformation beginning in the 1950s. First, vertical integration was central to the transformation of commercial poultry production on the Lower Maryland Eastern Shore. In the vertical integration process, a company controlled the ownership, production, and distribution of a product. In the commercial poultry instance, a vertically integrated company owned the eggs, chicks, feed, and the chickens and controlled their marketing before they reached the consumer. This process played an important role in the consolidation of commercial poultry production in Maryland.

The period of the 1950s to the 1990s brought about commercialization and consolidations that further transformed commercial poultry production. New techniques of feeding and housing chickens were invented during this period of transformation. These techniques improved the overall health and quality of the chickens. As commercial poultry gained prominence, concomitantly, African American involvement became more profound mostly in chicken catching. Their involvement in commercial poultry production shifted to poultry processing plant labor in the 1960s. In 1968, Frank Perdue purchased the old Swift and Company's run down building in Salisbury, renovated it, and established the Perdue poultry processing plant. Its establishment led many African Americans to gravitate towards the processing plant as assembly line workers. Beyond Salisbury, they were also employed in other poultry processing plants on the Lower Shore. They worked in all aspects of the labor supply-side in these plants and contributed to the regional economy of the Lower Maryland Eastern Shore. Also, their critical labor contributed to the Lower Shore being ranked among the top poultry producing counties both in Maryland and the nation during the 1980s and 1990s.[1]

VERTICAL INTEGRATION AND INTENSIFICATION OF COMMERCIAL POULTRY PRODUCTION

By definition, integration is when various parts of a whole are brought together under one whole. In the context of commercial poultry integration, the various components of commercial poultry—bird selection, breeding, egg laying, egg hatching, chick raising to broiler stage, slaughter, cleaning, packaging, transporting the chickens to the market, and public relations—were brought under the control of one company. Integration became a dominant factor in the industry during the 1950s after the large-scale commercialization of poultry production.[2] Essentially, the vertically integrated approach effectively eliminated many smaller scale poultry growers who did not have the resources to invest in superior methods of production to obtain high quality poultry.

According to Ensminger,

> Probably 95 per cent of the broilers of the U.S. are grown under some type of vertical integration or contractual arrangement. In the beginning, most commercial broiler production was by independent growers. They paid cash for everything and took all the profit. However, as margins became smaller, and flocks became larger, there was a need for more credit. At first, the local feed dealer was the source of credit. As the industry grew, feed dealers began to depend on feed manufacturers as a source of funds. Then to spread their risks, both feed dealers and feed manufacturers integrated vertically with hatcheries and processors.[3]

Having an integrated poultry production allowed producers who risked market unpredictability by investing in poultry, sometimes to the extent of borrowing money or exhausting their savings and even using their houses for collateral, to cut production costs while improving on product efficiencies in the sense that the financial risks were more spread out.[4] Since aspects of commercial poultry could be used for other things such as energy generation by burning poultry litter and poultry manure used as farm manure, vertical integration enabled producers to diversify their earning sources. If there was a high mortality rate among the chicks, the poultry farmers suffered huge losses. However, the losses were minimized when the poultry litters were converted and sold to crop farmers as manures. This way, the poultry farmers recouped some of their lost income from the mortality among the chicks.

Before vertical integration occurred on the Lower Maryland Eastern Shore, the practice was in place in Gainesville, Georgia, and in other parts of the country. An operator by the name of J. D. Jewell, a feed store operator in Gainesville, Georgia, bought a hatchery in 1940 for the purposes of consolidating the phases of commercial poultry production.[5] Along with the First National Bank of Gainesville, which loaned money to poultry operators, the stage was set for an integrated poultry industry. The bank established a client base of poultry operators. As the bank expanded its operations, poultry operators had a steady source of credit as well as loans to invest into their business. The resultant cycle enabled Jewell in particular to purchase a new feedmill by 1954.[6] Subsequently, the resultant development from Jewell's purchase of a new feedmill marked the early phase of vertical integration, a process that was further consolidated by the 1960s through the 1970s.

Jewell was a hands-on poultry producer in the sense that he was a frugal farmer who found innovative ways to use both wastes and by-products from his poultry processing plant in other areas of his commercial poultry operation. Through this practice, Jewell established a pattern in commercial poultry production whereby producers found alternative uses for their poultry wastes.

In other parts of the country, vertical integration followed the Georgia Model and led to the regional growth of commercial poultry. For instance, in Arkansas and Texas commercial poultry became an established enterprise after a series of failures in apple production led many farmers to seek out other sources of economic survival.[7] The abundance of and access to land particularly in the western part of Texas allowed poultry growers and potential growers an opportunity to become involved in the burgeoning enterprise. Thus, land economics made west Texas an early magnet for commercial poultry production in the Southwestern United States.[8]

The idea of transforming commercial poultry production on the Delmarva region came from J. Frank Gordy, a poultry specialist at the University of Delaware Extension Service. Gordy and the Extension Service promoted the idea of vertically integrated poultry, whereby the bigger commercial poultry companies oversaw the entire process of production from the grains fed to the chicks and broilers to their final destination—the dinner table.[9]

Having complete control was intended to ensure high quality poultry. The argument had been that poultry production passed through a variety of growers, many of whom had their own methods of production that might not have ensured high quality. The Extension Service further argued that an integrated approach was the preferred method because it helped weed out poor quality commercial poultry producers.[10] Supposedly, these inefficient producers slowed down commercial poultry's potential growth.

One significant area that helped transform commercial poultry production on the Delmarva region during the integration process was housing. Although other areas such as hatcheries and feedmills were important, housing was a major consideration because profits were made or lost in the poultry houses. In this sense, adequate housing arrangements included ensuring maximum comfort for the chickens by maintaining the room temperature at a certain degree Fahrenheit. The houses contained propane stoves which were used for brooding, and the chickens were fed through feeders and drank from automatic troughs.[11]

The process of poultry integration on the Lower Maryland Shore began in the 1950s and followed similar pattern of integration in other parts of the South.[12] Perdue Farms was an early leader of this effort. It built grain facilities, feedmills, hatcheries, soybean refinery, and processing plants that enabled the company to expand and consolidate its dominance of commercial poultry production. In the vertically integrated model, the company supplied the chicks, and the growers, in turn, purchased the feeds for their chickens from the company. Since the company had a stake in the overall transaction, it exercised considerable control and influence in the type of houses to raise the chickens.[13] Perdue saw integration as one way to maximize capacity as well as reduce production costs and gain a competitive advantage.[14]

In the 1950s and 1960s, the poultry houses were designed to be wider (40') with pole type construction, ridge vents, adjustable windows, or curtains than the prior decades.[15] Fans and insulation became regular features during this period. An added feature included multi-story houses where the growers lived on the top floor and the chickens were kept on the lower floor. On the Lower Maryland Shore in particular, the "housing density was 13,000 to 14,000 birds per house with .8 square feet per bird."[16] The housing arrangements were part of the vertical integration because the conditions under which the chickens were housed affected the overall quality of the chickens in terms of their appearance. In essence, vertical integration forced many small-scale farmers and growers out of business because of the prohibitive costs involved. It strengthened the competitive advantage of the bigger companies and effectively made it easier for commercial poultry production to be transformed.

As transformation continued in 1957, Perdue purchased the Townsend Farm in Salisbury and converted the land into multi-purpose commercial poultry use.[17] With the purchase of smaller poultry operations, Perdue was able to consolidate its dominance and gained a larger share of the local market than it had when it was just a father and son operation and focused more on egg production and marketing.[18] As part of its wider expansion strategy of integration and consolidation, Perdue entered into contractual agreements with local growers who raised chickens for the company. These growers, under the terms of the contract, were responsible for providing the housing and shouldered potential loses from the market uncertainties that resulted from fluctuations in demand and consumption.[19]

Most of the growers under contract for Perdue were whites who had easier access to loans.[20] Blacks, on their part, did not receive similar consideration. They had a harder time obtaining loans to engage in poultry growing. In general, whites were more likely to be extended credit opportunities than blacks, in part, because they had more collateral. It is plausible that Perdue preferred white growers because they presented less of a credit risk.[21] Furthermore, it is likely that there was a preference for white poultry growers because more whites owned poultry farms than blacks. In short, the social and economic outlook on the Lower Maryland Eastern Shore at the time favored whites more than it did for blacks.

To be considered a contract grower, the prospective grower had to provide evidence of financial wherewithal or credit worthiness. They received a flock of baby chicks from the poultry companies. Growers had four main systems under which they raised chickens for large companies:

> . . . the producer owns the houses and equipment, provides the labor, chicks, feed, fuel, disinfectants, etc., and receives all receipts. . . ; the producer owns the houses and equipment, and provides the labor, and a second party, possibly a feed dealer or hatcheryman, provides the chicks, feed, fuel, disinfectants, etc. When the broilers are sold, the items supplied by the third party are paid for out of undivided profits and the second party received from one-fourth to one-third of the balance of receipts. In cases where the receipts do not cover the expenses of the second party, the second party takes the loss; the second party pays the producer an agreed rental per stove or per 1,000 chicks for the use of the houses, equipment, and for the producer's labor. The second party receives all receipts, and [the fourth system] the producer owns the houses and equipment, provides chicks, feed, fuel, disinfectants, etc., and a second party provides labor only. The second party may get a percentage of the receipts after the cost of chicks, feed, fuel, disinfectants, etc. are paid for. However, in other cases the second party is little different from a hired man and may be paid an agreed amount per stove, per 1,000 chicks, or per month for his labor. The producers usually build living quarters for their labor, oftentimes on the second floor over a section of the poultry houses. The laborer is often given bonuses and other incentives for exercising a high degree of poultry husbandry.[22]

Contract growing was both an elaborate and expensive endeavor.[23] The high cost of maintaining the terms that were attached to the contracts shaped the commercial poultry ownership class in the sense that few people were able to meet the terms. There were many local residents who wished to become contract growers but could not because they did not meet the conditions attached to any of the four systems that were articulated by Bausman. Under the contractual systems, the large companies had an advantage because they controlled how many chicks, how much feed, and how much disinfectant that the potential growers received. Not only that, the large companies had the ability to charge the growers any amount that ensured that they recouped monies already invested.

Many of the associated costs were passed on to the growers as the companies required the growers to abide by high standards of quality. The expensive nature of commercial poultry contracting meant that most blacks' access to it was curtailed because they did not have the needed capital. In some instances, the mostly white prospective growers qualified for start-up loans from the area lending institutions. Such credit worthiness was a disqualifier for blacks because many of them did not have the same credit opportunities as whites. Consequently, poultry giants like Mountaire, Perdue, and Tyson usually extended growers' contracts to credit-worthy whites.[24]

Although whites dominated the growers' ranks, black labor became evident when the chicks reached maturity as broilers.[25] Blacks largely were the ones that caught these chickens at the chicken houses. There were poor whites that also caught chickens, but their numbers were negligible compared to black catchers.[26] On their part, whites mostly drove the trucks that transported the broilers to the processing plants. Upon arriving at these plants, blacks unloaded the broilers from the trucks and delivered them to the slaughter areas of the plants.

Vertical integration on the Lower Maryland Shore proved disastrous for some independent growers and producers as they were effectively pushed them out of commercial poultry production. Consequently, these independent growers from the region founded the Eastern Shore Poultry Growers' Exchange in 1952 in Selbyville, Delaware. The main purpose of this Exchange was to counter the gradual monopolization of chicken feed and hatcheries. The dominant producers of chicken feed and hatcheries on the Lower Maryland Shore at the time were Pillsbury, Quaker Oats, and Purina Foods.[27] As Perdue began its expansion, it drove out these chicken feed companies so that by the late 1950s, Perdue established its own chicken feed operations.[28] The company's rise as the preeminent chicken feed producer, coupled with its integrated approach stifled other competitors; hence, it's regional dominance in commercial poultry production.[29]

Perdue and other bigger companies' dominance motivated independent poultry growers to sell and market their own products at publicly held auctions between 1952 and 1969 to try and reduce the influence of the big companies. The big companies' terms for contract growers were not as beneficial for the growers as they were for the companies. In some instances, the growers were forced into borrowing more money to meet the companies' specifications such as how long and wide and how properly ventilated the poultry houses should be to fit the companies' relative preferences for quality. These specifications were costly to the growers, some of whom went into more debt because they had to borrow more money to meet the terms of their contracts. The auctions were also demonstrations of independence from the big companies.[30] They challenged the companies' domination of commercial poultry production. For many, poultry was their livelihood as many growers took out loans from commercial lending institutions and purchased lands, buildings, and equipment to meet the growing demand for poultry.[31]

Despite efforts by independent commercial poultry producers to assert themselves as competitors to the bigger companies through the poultry exchange auctions, vertical integration continued from the 1960s through the 1990s. By the 1970s, Perdue continued its integration mainly through advertising. The company used its vast network of customers that it established through its radio, newspaper, and television advertising, which had grossed $80 million in sales for the company by 1972.[32] In the 1980s, Perdue continued its integration of commercial poultry production with the acquisition of Shenandoah Valley Poultry Company and Shenandoah Farms, which further expanded its reach and influence beyond Maryland and into Washington, Indiana.[33] These acquisitions added to Perdue's control and ownership of hatcheries, feedmills, and processing plants.[34] The expansions strengthened Perdue Farms and made it more than just a local and regional commercial poultry producer. It became a major national commercial poultry integrator. It grew larger and acquired new operations including existing hatcheries, feedmills, and processing plants that helped expand sales. Perdue Farms pursued an innovative strategy where it expanded without constructing new infrastructure to accommodate the new acquisitions.

On the Lower Maryland Shore, integrated poultry affected African Americans, but in different ways than it affected white poultry farmers. In the first place, few African Americans owned large acre poultry farms that could be used for various purposes or expanded to accommodate any one of the components of integrated poultry. African Americans did not wield the requisite power to influence the outcomes of integrated poultry because of the lack of large poultry farm ownership.

COMMERCIALIZATION AND CONSOLIDATION OF COMMERCIAL POULTRY PRODUCTION, 1950s TO 1970s

During the decades of the 1950s through 1970s, several inventions were developed, which further transformed commercial poultry production. For example, one invention "nipple drinkers helped improve housing conditions for the birds as they are being grown . . . and the increased use of tunnel ventilation also [made] for a better environment for the chicken houses for the birds and for the people who work in the [chicken] houses."[35] In particular, the nipple drinker helped minimize debris from getting into the water compared to an open drinking container as was the case before the 1950s. The cleaner drinking water reduced the chickens' contraction of water-borne diseases that would have been present in open containers. The tunnel ventilation also reduced air-borne bacteria present in unfiltered airflow. This method was beneficial for poultry house workers because of the enhanced air quality which was devoid of fecal matters present in chicken feed. Lower Eastern Shore commercial poultry producers adopted the tunnel ventilation method to further distinguish the quality of their chickens from those produced in other places.[36] This method of poultry management furthered the "Delmarvalous" depiction of chickens produced on the Lower Maryland Eastern Shore.[37]

As commercial poultry became a large-scale enterprise in the 1950s, transportation became more of a consideration. The Lower Shore producers needed to export their poultry products to other parts of the State and region. However, in terms of transportation, the Lower Shore was isolated from the rest of the State. In fact, traveling to the Shore was either by boat or some other form of sea-based transportation. Traveling by road required that you go through Delaware to reach the Lower Maryland Eastern Shore.[38] Consequently, elected officials thought up a plan to connect the region with the rest of the State and hoped to break the transportation isolation.

The governor of Maryland at the time, Theodore McKeldin, a Republican and former mayor of Baltimore City, was a strong advocate of transportation and economic development. He especially pushed for transportation connectivity throughout the State. The governor prevailed in the debate that followed in the Maryland General Assembly. The major provision of his highway master plan called for a series of road constructions that linked all parts of Maryland. This was not only accomplished through inter-state highways but also included rural roads in all parts of the state.[39]

The end result of this spirited debate was the Chesapeake Bay Bridge, which opened in 1952. With the bridge in place, central Maryland became connected with the Eastern Shore. The bridge was crucial to Perdue and the other commercial poultry producers' ability to market poultry outside of the Lower Shore. Perdue Farms took advantage of this bridge and established a transportation fleet that moved its poultry products to other parts of the state with greater frequency. Its ability to utilize the inter-state transportation links expanded its customer base beyond Maryland and into other parts of the East Coast including Philadelphia, New York, and Baltimore. The multiplier effect of the increased demand for poultry spurred by improved transportation led to the further expansion of poultry production. The bridge also spurred the growth of the regional economy as commercial activities flourished.[40] The new bridge was essential to the viability of commercial poultry on the Lower Eastern Shore. In this instance, transportation positioned the region for further advancement and growth of commercial poultry production.

During the 1950s, there were indicators that showed the sustained viability of commercial poultry on the Lower Shore. Maryland in general was ranked sixth in the nation in commercial hatchery operations, ahead of Delaware. Perdue in particular was a major contributor to this national recognition in part due to its expanding operations including contracting with local growers.[41] The commercial poultry industry on the Lower Shore in the 1950s was profitable. For instance, in 1954, a total of 981 poultry farms in the three counties reported selling $38,849,891 million worth of poultry. In 1959, 1,274 poultry farms (an increase of 293 farms) in the same counties reported selling $23,118,335 million worth of poultry, a decrease of more than $15.7 million.

Table 4.1. Number of Poultry Farms on the Lower Maryland Eastern Shore, 1950–1964

	Year	*State*	*Somerset*	*Wicomico*	*Worcester*
Total	1950	3,462	251	465	497
	1954	2,796	251	457	273
	1959	NA	259	508	507
	1964	NA	261	421	346

NA = Not available

Source: U.S. Census of Agriculture, 1954, Volume 1, Part 14, County Table 3, 165; Maryland Census of Agriculture, 1959 and 1964: 70–78.

Table 4.2. Poultry Sold and Farms Reporting on the Lower Maryland Eastern Shore, 1954–1959

	Year	*Somerset*	*Wicomico*	*Worcester*
Poultry sold ($)	1954	8,629,100	16,412,125	13,808,666
	1959	4,839,955	10,291,739	7,986,641
Farms reporting	1954	404	813	709
	1959	347	760	625

Source: Ibid, Table 10, 145.

Table 4.3. Number of Broilers Sold on the Lower Maryland Eastern Shore, 1954

	Somerset	*Wicomico*	*Worcester*
Farms reporting	270	431	290
Number of broilers	6,988,860	14,887,544	11,470,942

Source: U.S. Census of Agriculture, 1954, Volume III, Table 10. [By 1954, Maryland was 5th in the number of broilers sold, accounting for 5.8% of the total U.S. broiler production]; Ibid, Table 9.

The decline in the number of farms reporting could have been as a result of financial difficulties necessitated by the fluctuations in the poultry market. Many farmers lost their farms because they were unable to pay back farm-related loans. In some cases, farmers used their farms as collateral to obtain loans, which placed their farms in jeopardy in the event that they could not pay back their loans.[42] Also, the shift from an agricultural economy to urban industrial factories in the 1950s, as well as the expansion of agro-business, accounted for another reason some farmers lost their farms.

By the 1960s, Perdue Farms' presence in the Salisbury community was firmly entrenched with the processing plant. Its presence replaced another long-standing institution in the black community in terms of employment, Swift and Company, which at the time was a local institution with a large black work force. The company had been a constant presence in the black community. When Frank Perdue renovated and converted the building into a more modern poultry processing plant in 1968, the plant became a key player in the regional economy of the Lower Maryland Eastern Shore.[43] One local resident stated:

> The plant became a magnet for black workers just as the old Swift and Company was prior to its sale. Hundreds of blacks got jobs at the new plant. For many blacks, the plant was in the community and therefore convenient. The workers did not have to travel a long distance to get to work. In fact, many of them walked to work. It was new and seemed to provide different sets of economic possibilities for local blacks as the old Swift and Company was gradually fading away from the community. The odors from the plant were strong and caused many of the black convenience stores, restaurants, and nightclubs, all of which were located in the same area, to move. Black Salisbury was a thriving community but the smell from the plant was just too strong. As a result, black businesses lost patrons. The smell drove them away. When the stores closed down, Perdue quickly stepped in and became the major force and employer in the Salisbury black community. It provided employment for the local residents many of whom had worked for the businesses that closed or relocated.[44]

Commercial poultry production further developed and expanded in the 1970s. Perdue Farms continued its quest for dominance by launching new ideas and research and expanded its operation throughout the mid-Atlantic region in response to increased demand by establishing "breeding and genetic research programs" to enhance both the quality and quantity of its poultry meat.[45] These programs furthered strengthened and positioned Perdue Farms as the dominant commercial poultry producer on the Lower Maryland Shore.

Perdue Farms' foray into the genetic research field was a major boost. For the company, appearance and the size of its chicken was indicative of a healthy breed. In 1974, the company introduced the PERDUE Oven Stuffer

Roaster and distinguished its chicken meat from other competitors by cross-breeding Perdue's chickens to produce much meatier chickens.[46] To further exploit this new breed of meatier chicken, Frank Perdue launched a national television campaign to aggressively advertise its new product and thereafter ushered in a new era of commercial poultry populism. The television commercials gave Frank Perdue an advantage in the court of public consumption opinion over his competitors.[47]

Perdue's poultry populism was not restricted to his operations. His television outreach commercials helped propel Maryland as a whole, which in 1970 was ranked among the top six commercial poultry producing States in the nation, to greater heights. By the end of the 1970s, the value of Maryland poultry had exceeded $300 million and each of the counties on the Lower Shore was ranked among the top poultry producing counties in the nation.[48] In a sense, the purchase and renovation of the old Swift & Company plant perhaps played the biggest role in Perdue Farms' dominance of commercial poultry production not just on the Lower Maryland Eastern Shore, but also on the Delmarva region during the decade of the 1970s. As the commercial center on the Shore, Salisbury attracted buyers from distant places. This served as a motivator for the Salisbury plant to produce commercial poultry at a higher rate, which it did at 12,000 broilers an hour between 1967 and 1972.[49] Thus, the Salisbury plant became the preeminent poultry processing operation on the Maryland side of the Delmarva.

COMMERCIALIZATION AND CONSOLIDATION OF COMMERCIAL POULTRY PRODUCTION, 1980s TO 1990s

Commercial poultry production on the Lower Maryland Eastern Shore went through further consolidations and development in the 1980s and 1990s as Perdue, Mountaire, and Tyson Foods acquired small to mid-size operations such as Holly Farms and Townsend Foods. These acquisitions, mergers, and expansions further consolidated Perdue's already well established influence

Table 4.4. Rankings of Lower Maryland Eastern Shore Counties with Chicken Sold in the Nation (1978)

County	*Broilers Sold*	*Rank*
Somerset	33,008,942	19th
Wicomico	64,033,579	4th
Worcester	55,031,424	6th

Source: Delmarva Poultry Industry, Inc., (DPI Archives) Georgetown, Delaware.

as the economic power in the region in terms of being a source of employment for many local residents.[50] Consequent to these acquisitions, expansions, and consolidations Perdue gained a competitive advantage over the smaller and independent commercial poultry producers that still operated on the Lower Shore.

Although Mountaire and Tyson Foods were competitors of Perdue Farms and operated on the Lower Eastern Shore, their impacts and roles were limited or confined to commercial poultry transactions in the sense that they were not rooted in the communities and not as actively and extensively engaged in activities such as constructing schools or recreational centers specifically for the local communities. In Delaware and Arkansas, Mountaire and Tyson Foods both have invested in communities throughout these two states because they are headquartered there and their operations were rooted in these two states. In comparison, Arthur and Frank Perdue were born on the Lower Eastern and were therefore wholly invested in their communities. Both were involved in charitable causes and extensive philanthropic works such as donating monies for scholarships and poultry-related research on various issues that were important to the Lower Eastern Shore. They donated money for various other local causes and helped build schools and recreational facilities.[51]

By the 1980s, the value of commercial poultry continued to flourish in Maryland. Throughout the decade, the average yearly total value of commercial poultry in the State was $367.3 million.[52] Wicomico County was ranked fifth in broiler production and sales in the United States in 1982 and 1987(as shown in Table 4.5). These rankings translated to employment and revenues for the local and regional economies.

Table 4.5. Rankings of Lower Maryland Eastern Shore Counties with Chickens Sold in the Nation, 1982 and 1987

County	*Broilers Sold*	*Rank*
Somerset		
1982	39,443,031	13th
1987	45,671,603	17th
Wicomico		
1982	68,466,740	5th
1987	72,735,534	5th
Worcester		
1982	62,428,517	7th
1987	67,860,514	8th

Source: [Brochure] Delmarva Poultry Industry, Inc., Georgetown, Delaware.

By 1982, in a bid to enhance the regional economic viability, lending institutions such as Farmers Home Administration, an agency of the United States Department of Agriculture, was set up to encourage and assist farmers with their business-related issues. Many local residents were encouraged to go into the poultry business enterprise as growers.[53] There was also a Farm Credit Association, as well as a self-help cooperative of farmers and lenders in the three counties, which sought to assist members with obtaining necessary credit with low interest. They were crucial avenues for local poultry growers.[54] In a way, this association minimized the adverse effects of large-scale poultry companies on the independent small-scale poultry producers in the sense that the large-scale companies had the means and resources to obtain financing to expand their business operations. On the other hand, the small-scale companies had limited resources to sustain the terms of a large credit line from a commercial lending institution.

While Frank Perdue was promoting his "high quality" poultry meat on national television, he was also recruiting potential growers. The television commercial elevated Perdue's standing in the meat industry. The company aggressively promoted the idea that it, "[Perdue] offers you the opportunity to start your own business."[55] As a result, more potential growers flocked to the company's invitation for business opportunity. This became a part of a strategy to corner the regional market and dominate his main competitors, Allen, Mountaire and Tyson Foods. Given their weak financial position, very few African Americans were in a position to become contract growers for Perdue. In the local sense, the growers were part of the "favored" class within the local economic power structure. In this case and coupled with the racial dynamics on the Lower Shore at the time, the growers were overwhelmingly white.[56]

Actually, Perdue's slogan was not really what it seemed. In view of the vertical integration that had taken place, the company owned much of what a grower needed to successfully operate a poultry business with the exception of housing, which was owned by the grower. Thus, in reality, a grower did not really "own" the business. He or she was operating the business on behalf of Perdue. The burden on growers was such that they would have to enlarge the size of their poultry flocks to make a profit, supposing there was minimal poultry mortality. They had to first meet their contractual obligation to Perdue. They had to borrow money from a lending institution and purchase flocks and feed grains from Perdue. All these obligations increased their debt. It was a promotional advertisement that worked to Perdue's advantage, as the company was able to increase its contract growers' ranks among whites.[57] Contract growers on their part built their poultry houses in accordance with the poultry companies' specifications. The exorbitant cost of the poultry houses, at the time estimated to be about $250,000, caused many growers to go further into debt.[58] The growers competed against "Big Chicken" as it dominated commercial poultry production.[59]

By the 1980s, there was tension within the growers' ranks. "Big Chicken" had effectively rendered small and independent growers poor as they struggled to make a profit. "Big Chicken" became a factor because "a lot of poultry farmers consider themselves as serfs on their own land. There is no control by the farmer over anything. You either take it or leave, and that's not right."[60] The contracting approach enabled the big companies to dominate all aspects of commercial poultry production. Since the three Lower Shore counties were the epicenter of Maryland's axis of poultry, the growers had little option but to buy the corn and grains for their flock from the big companies. As part of the terms of agreement, the growers purchased all their feed from the companies.[61] As the growers became "dependent" on the big companies, commercial poultry production on the Lower Shore was systematically consolidated to the extent that there were fewer than five small and independent commercial poultry operators on the entire Lower Shore by the end of the decade.[62]

During the 1990s, Perdue Farms continued its development and expansion through acquisitions.[63] The company purchased Showell Farms, Inc. in Worcester County, a commercial poultry processor, in 1995.[64] This acquisition expanded and improved Perdue's profit margins. Although Perdue was not the only commercial poultry producer on the Lower Maryland Eastern Shore, it became the major force through its vertical integration of commercial poultry production on the Maryland side of the Delmarva region.

In effect, the value of commercial poultry compared to the 1980s rose by $94 million. In the decade of the 1990s, a record number of poultry was sold on the Lower Shore (as shown in Table 4.6). Indeed, the 1990s, the average yearly value of commercial poultry in Maryland during the decade was $461.3 million.[65] The large volume of poultry sold on the Lower Shore demonstrated commercial poultry's importance to both the local and regional economies. Although Perdue Farms was the dominant poultry producer, it faced competition from other poultry producers. Tyson Foods, for example, which operated on the Lower Shore in the 1990s, earned 83 percent of its annual revenue in 1997 from commercial poultry (consumer poultry).[66] The

Table 4.6. Amount of Poultry Sold on the Lower Maryland Eastern Shore, 1992

State	257,209,663
Somerset	48,523,355
Wicomico	76,497,668
Worcester	57,407,806

Source: Census of Agriculture, 1992—County Data, Table 16, Maryland, 235.

high percentage of Tyson Foods' annual revenues from commercial poultry demonstrated competition in the sense that the company had to slaughter and process a high number of birds to be able to keep up with Perdue Farms.

A combination of factors contributed to the overall economic prospects of commercial poultry on the Lower Maryland Eastern Shore. Improved growing techniques such as injecting chickens with antibiotic growth hormones and mixed chicken feed, were part of the changes that aided in the increased production output. At the same time that the total value of commercial poultry increased, the number of chickens produced also increased. By 1999, growers, according to the Delmarva Poultry Industry, numbered 2,531 (as shown in Table 4.7).

AFRICAN AMERICANS AND COMMERCIAL POULTRY PRODUCTION

The commercialization of poultry on the Lower Maryland Eastern Shore brought about an increased African American participation and involvement.[67] There were other sectors of the local economy on the Lower Shore such as vegetables and fruits, but these sectors were seasonal and often the earned wages were insufficient to care for a large family. In comparison, commercial poultry was year-round and the income was steady. Blacks worked in both the crop and produce sectors of the local economy as well as in the commercial poultry production. However, the social conditions at the time such as racism, discrimination, and segregation restricted blacks to mostly labor-type jobs. As commercial poultry became sustained, proportionately, more blacks became involved as well.

The development of employment opportunities in the industrial and tourism sectors on the Lower Shore, led many blacks to migrate from rural areas to seek new opportunities in these sectors. Some blacks emigrated out of the rural areas to towns and cities to take advantage of the "industrialization" of Salisbury (Wicomico County) and the tourism of Ocean City (Worcester County).[68] Their out-migration had minimal impact on their lifestyle because many blacks found themselves working in the processing plants.

Table 4.7. Delmarva Regional Poultry Statistics, 1999

Number of Poultry Houses	5,816
Poultry Houses Capacity	131,000,000
Poultry Growers	2,531
Poultry Company Employees	14,000

Source: [Brochure] Delmarva Poultry Industry, Inc., Georgetown, Delaware.

According to the U.S. Census of Population for the counties (as shown in Table 4.8), African American population steadily increased, in some cases slightly from 1950 through 1990. By 1960, black population in the three counties grew, albeit at a slow pace. In Wicomico County, the black population increased by 2,652 people. The increase in population is instructive because the county, especially the city of Salisbury, provided the best economic opportunity when compared to other available economic activities for blacks. Also, the employment trend positioned the county as the economic hub on the Shore mainly because of the Port of Salisbury. The city also gained population because many blacks who left Somerset and Worcester Counties were part of the intra-regional migratory trend from rural to urban areas.

The trend in increased black population continued in Wicomico and Worcester Counties in the 1970s with the exception of Somerset County, which lost 247 residents. It is probable that the slight decrease in population was attributable to better economic prospects in Salisbury or even Ocean City both of which boasted diversified industries. In 1980, both Wicomico and Worcester Counties gained an increase in their black population compared to 1970 while Somerset lost more than 400 of its black residents. As the most rural of the three counties, perhaps blacks left to take advantage of the industrial urbanization of Wicomico County in particular.

By the 1990s, Somerset and Wicomico Counties' black populations increased while Worcester County's black population declined. The varied black population shifts in these counties from the 1950s to the 1990s might be because of competition from Hispanics who first registered their presence in Wicomico County in 1970. The Hispanics who first arrived in the 1970s worked in the poultry processing plants and this created a competition with black workers.[69] In a way, this competitiveness motivated black workers in the sense that the relative stability of their employment served as an incentive to work even harder. After all, many of the Hispanic workers were non-citizens and undocumented immigrants. It seemed that their presence at the processing plants gave black workers an added advantage because the poultry companies had to weigh violating immigration laws with heavy fines or retain their black workers even when they paid black workers more than the lower paid Hispanic workers.

During the period of the 1950s and 1960s, the Lower Maryland Eastern Shore was still a tense region. Blacks battled economic inequality and lack of civil rights. Politically, they fought against the local white power structure which defined the parameters within which blacks participated in the political process as whites held virtually all the locally elected offices. It was not until the 1960s that local leaders such as Gloria Richardson and Enez Stafford Grubb and outsiders such as H. Rap Brown brought defiant

Table 4.8. Racial Breakdown of Census of the Lower Maryland Eastern Shore, 1950–1990

Race	*State*	*Somerset*	*Wicomico*	*Worcester*
1950				
Total	2,343,001	20,745	39,641	23,148
Whites	1,954,975	13,904	30,795	16,046
(%)	83.4%	67.0%	77.95%	69.3%
Blacks	385,972	7,326	8,372	7,094
(%)	16.5%	35.3%	21.1%	30.6%
1960				
Total	3,100,689	19,623	49,050	23,733
Whites	2,573,919	12,263	38,026	15,585
(%)	83.0%	62.5%	77.5%	65.7%
Blacks	518,410	7,360	11,024	8,148
(%)	16.7%	37.5%	22.5%	34.3%
1970				
Total	3,922,399	18,924	54,236	24,442
Whites	3,193,021	11,800	42,636	16,397
(%)	81.4%	62.4%	78.6%	67.1%
Blacks	701,341	7,113	11,512	8,017
(%)	17.9%	37.6%	21.2%	32.8%
Hispanic	52,974	—	443	—
(%)	1.3%	—	0.8%	—
1980				
Total	4,216,975	19,188	64,540	30,889
Whites	3,158,838	12,433	49,679	22,593
(%)	74.9%	64.8%	77.4%	73.1%
Blacks	958,150	6,639	14,085	8,100
(%)	22.7%	34.6%	21.8%	26.2%
Latino	64,746	125	408	235
(%)	1.5%	0.7%	0.6%	0.8%
1990				
Total	4,780,743	24,440	74,339	35,028
Whites	3,393,964	14,282	56,755	27,253
(%)	71.0%	60.9%	76.3%	77.8%
Blacks	1,189,899	8,943	16,573	7,467
(%)	24.9%	38.2%	22.3%	21.3%
Latino	125,102	229	610	275
(%)	2.5%	1.0%	0.8%	0.8%

Source: U.S. Census Bureau; Maryland Populations, Maryland State Data Center, Department of Planning U.S. Bureau of the Census, 1950 to 1990.

activism to Cambridge and challenged the white power structure. Although his activities were mainly centered in Cambridge, Dorchester County, Brown attracted the attention of the other Lower Shore counties' white establishment with fiery speeches about economic oppression against blacks. His radicalism inspired local black activists who re-assessed their strategies for fighting white domination.[70]

Although commercial poultry workers would have benefited from the social and economic activism of Richardson, Grubb, and Brown, the workers did not fully and enthusiastically join the protests for fear of losing their jobs either in the chicken houses or in the processing plants. The intimidation factor was a real threat that discouraged workers from joining the protests. Local white authorities used the label "trouble makers" to brand critics of the local power structure. Had the workers participated in the protests and were then branded as "trouble makers," their means of livelihoods would have been severely affected by such stigmatization.[71]

The 1950s and 1960s were turbulent periods not just in civil rights and social arenas, but also in access to education. Lack of access to education was a factor in commercial poultry production on the Lower Eastern Shore. In the 1950s, schools on the Lower Shore were still segregated. However, with the landmark Supreme Court case, *Brown v. Board of Education* ruling that struck down the "separate but equal" practice, the Shore implemented the ruling of the Court. Despite the ruling, local whites found other ways such as under funding black schools to frustrate the intent of the ruling. In fact, from the 1950s through the 1970s, the idea of integrated public accommodations for blacks on the Lower Shore counties was still far-fetched. Racial tensions persisted. Blacks still ate at, and drank from separate public facilities. They could not socialize with their white counterparts.

Educationally, blacks were underrepresented in the schools. In 1950, education for blacks in the three counties remained problematic. For example, black student enrollment in grades K through 12 was the lowest in Wicomico County at 26% of total student enrollment compared to 74% of white student total enrollment. In Worcester County, black student total enrollment was 36.8% and white students were 63.2% of the total student enrollment. Somerset County had a higher enrollment with blacks at 40% of the total student enrollment while white students constituted 60% of the total student enrollment.

According to the Eighty-Fourth annual report of the State Board of Education of Maryland in 1950, some blacks completed high school only in Wicomico County while there were no black high school completions in Somerset and Worcester Counties.[72] One reason was that Wicomico County was more diverse in terms of economic opportunities and blacks who did not want to work in the poultry industry had more employment prospects in

Table 4.9. Condition of Education on the Lower Maryland Eastern Shore, 1950*

	Somerset		*Wicomico*		*Worcester*	
School Enrollment+	*Whites*	*Blacks*	*Whites*	*Blacks*	*Whites*	*Blacks*
Total:	2,141	1,443	4,522	1,617	2,454	1,426
Percent of total students enrolled in school	60%	40%	74%	26%	63.2%	36.8%
School completion**	4	—	188	47	4	—

*Local school boards used different variables to capture certain categories of educational statistics. For example, the variables that were captured in the 1940 census were eliminated in the 1950 census. Therefore the variables for the 1960 educational census differed from the 1950 census perhaps to reflect court rulings concerning school desegregation.
+ Includes K–12 enrollment in public high schools.
**High school graduates only, 1949-50, Ibid, Table 24.

Source: Eighty-Fourth Annual Report, State Board of Education of Maryland, 1950; Table 16 and 17.

other areas compared to Somerset or Worcester Counties where there were fewer economic diversities and opportunities. A "protracted" black poultry labor force emerged out of the lack of black school completion because without education, blacks were relatively less likely to work in higher-end jobs requiring demonstrable skills of a high school educational background.[73] Consequently, commercial poultry producers benefited from blacks' relative lack of education.

Table 4.10. Condition of Education on the Lower Maryland Eastern Shore, 1960*

	Somerset	*Wicomico*	*Worcester*
School enrollment			
Number enrolled in school by age			
5 and 6 years old	351	1,041	517
7 to 13 years old	2,379	6,515	3,137
25 to 34 years old	59	69	77
Percent enrolled in school by age			
5 and 6 years old	47.1%	47.0%	46.1%
7 to 13 years old	88.5%	97.9%	98.4%
25 to 34 years old	2.9%	1.1%	2.6%
School completion			
Males, 25 years old and over	5,406	13,393	6,604
No school years completed	182	297	243
Females, 25 years old and over	5,794	14,738	7,046
No school years completed	141	199	179

*The 1960 annual report by the State Board of Education of Maryland did not contain corresponding variables to the 1950 report.

Source: Census of Population, 1960; Maryland, General Social and Economic Characteristics, 22-151

When Perdue Farms took its place in the black community, a partnership of sorts developed that spanned several generations of blacks. Blacks worked in commercial poultry in various capacities. Poultry, therefore, was an integral part of black life on the Lower Shore during the period of large-scale commercialization of poultry. The poultry processing plant transformed the black community in the sense that poultry processing employment was available to blacks who were looking for work.[74] These jobs, while low paid, helped blacks provide for the family's basic necessities, such as housing and healthcare. In turn, the black community transformed poultry in the sense that their "reliable" labor was crucial to commercial poultry's profitability.[75] For many blacks, it was a family affair as there was a job at the plant for those who needed employment. But the jobs at the plant were prone to injuries. Studies showed that "one in five poultry workers is injured on the job" as workers were required to operate automated "mechanical killing and defeathering machines."[76] Many of the workers had minimal skill to operate the required machines. Consequently, they were prone to accidents on the job. Notwithstanding the risks, blacks continued to work at the processing plants.

Blacks continued their poultry labor even as the workers' ranks grew in the 1960s. Perdue Farms would not avail the specifics of its employees' racial statistics, but several of the black poultry workers interviewed indicated that blacks dominated the production line at the Salisbury plant since the plant went into operation in 1968.[77] The interviewees were either high school graduates or dropouts. Many blacks left before graduating from high school because they needed to work and help support their families. From their responses and the available educational data, the educational gap was a major factor in black involvement in commercial poultry production on the Lower Maryland Eastern Shore.

In effect, there was a correlation between education and the labor force of large-scale commercial poultry production on the Lower Maryland Eastern Shore. Perhaps if there had not been racism, discrimination, and segregation all of which led to the limited access to education for blacks, and if the counties had provided equal funding for the schools, blacks may not have dominated the labor ranks in commercial poultry production. The lack of educational achievement had a confining effect on blacks in the sense that potential employers did not regard black high school graduates as having similar skills as white graduates.

There was a division of labor between blacks and whites within the commercial poultry plants during the 1960s. Blacks were responsible for eviscerating the broilers in extremely hot water. It was not a common practice for black workers to wear protective gear that shielded them from being burned. They had minimum protection from the dangers of extremely hot water.

Blacks also worked as cutters, which required the removal of the broilers' intestines. By the time whites got involved mainly as supervisors blacks had done much of the grunt work.

In Worcester County, blacks had limited choice about where they worked and the conditions under which they worked. For example, because the Showell Poultry plant was located some distance away from the black community, blacks required transportation to get to their jobs. Some of the black workers lived in Pocomoke City, a distance of about 5 to 10 miles. In some cases, they carpooled and contributed a small amount of money to the driver. In the event the companies had transportation, the workers were charged for the transportation just as they were charged for things such as work gloves, aprons, boots, and other work-related items. Although the workers were supplied with their first set of gloves and aprons, they were required to pay for subsequent supplies when they needed them. The charges were deducted from their paychecks. The practice of deducting charges from the workers' paychecks created additional hardship, as their take home pay further dwindled. Notwithstanding, they persevered.[78] Although Hispanics began arriving on the Shore during the 1970s, African Americans dominated the commercial poultry production labor ranks mainly in chicken catching and processing plant work.

Their involvement in the commercialization of poultry on the Lower Maryland Eastern Shore did not change much in the 1980s. However, the influx of Latino workers led to a slight re-alignment in the sense that the rural to urban shift enabled local black residents to out-migrate to Ocean City and participate in the tourist business from which they had been excluded during the heyday of "official" and unofficial segregation.[79] Generally, however, blacks worked in Ocean City but lived in the rural section and away from the expensive ocean front properties.

While blacks on the Lower Maryland Eastern Shore made some gains during the decade of the 1990s, many still worked in commercial poultry production. Leadership opportunities, though small, became available as they rose to become crew leaders and line and shift supervisors. For black workers, it signaled career advancement for those who had worked in the plants for long periods of time.

CONCLUSION

The 1950s through the 1990s was the period of sustained expansion of commercial poultry production on the Lower Maryland Eastern Shore. It was also a period of innovation that further solidified poultry's status in Maryland's economy as the top source of income for Maryland's farmers. Equally, the construction of the Chesapeake Bay Bridge raised the importance of commercial poultry to the region's economic viability as commercial poultry producers were able to transport their products beyond the Shore. The impact of commercial poultry production on the region's economy remained significant during the period of the 1950s through the 1990s.

Perhaps, more groundbreaking transformation was the introduction of the vertical integration model in commercial poultry production. Although it changed commercial poultry production, it also impacted the economic conditions of African Americans on the Shore. Commercial poultry did ensure income stability; however, blacks were frozen out of the ownership and contract growers' ranks because of vertical integration. In the end, blacks were the most impacted because they did not have the wherewithal to meet the demands and expectations of the integrators. Although vertical integration sought to improve the quality of poultry meat to the consumer because of superior cross-breeding and other innovative methods, it ultimately had a negative impact on small and independent poultry producers. On their part, African Americans were essentially relegated to spectator status in the sense that their involvement did not include ownership within the integrated process. Rather, their participation was largely confined to the labor supply-side of commercial poultry production.

The process of development and consolidation furthered the prospects of commercial poultry. New and improved methods of raising, housing, and feeding chickens were developed during the large-scale commercialization phase and these methods combined to improve the overall supply, demand, and consumption of poultry. Despite the performance of commercial poultry in terms of the total value of production throughout the large-scale commercialization phase, poultry workers hardly benefited from the huge profits that accrued to the big companies. In spite such profits, the workers still earned low wages.[80]

The development and consolidation of commercial poultry production on the Lower Maryland Eastern Shore created a permanent labor class, whose involvement was largely responsible for the expansion of the industry. Importantly, this development and consolidation was also aided by the low level of educational achievement of the local workforce. The prevailing social conditions, including racism, discrimination, and segregation, gave rise to lack of quality educational attainment that would have provided local residents with more occupational options. The inability of many blacks on the Lower Maryland Eastern Shore to benefit from education and training in other fields enabled commercial poultry to thrive and subsequently led many local residents into the commercial poultry production labor force.

NOTES

1. Delmarva Poultry Industry, Inc., (Archives) Georgetown, Delaware.

2. William L. Henson, "The U.S. Broiler Industry: Past and Present Status, Practices, and Costs," *Agricultural Economics and Research Service*, 149 (May, 1980), 26, passim.

3. M. E. Ensminger, *Poultry Science*, First Edition (Danville, Illinois: The Inter-State Printers and Publishers, Inc., 1971), 163.

4. David M. Theno, "The U.S. Poultry Industry: Past/Present/Future," *Reciprocal Meat Conference Proceedings*, Volume 37 (1984), 3.

5. J. Frank Gordy, "Broilers: Fifty Year Old Meat Industry Presents Picture of Specialization," In John L. Skinner, et al. (ed.), *American Poultry History, 1823–1973* (Madison, Wisconsin: American Printing and Publishing, Inc., 1974), 382–384.

6. Ibid, 383.

7. Ibid.

8. Skinner, 383.

9. Mitzi Perdue, *Frank Perdue: Fifty Years of Building on a Solid Foundation* (Salisbury, Maryland: The Arthur W. Perdue Foundation, 1989), 63–73.

10. Roger Horowitz, "'Be Loyal to Your Industry': J. Frank Gordy, Jr., the Cooperative Extension Service, and the Making of a Business Community in the Delmarva Poultry Industry, 1945–1970," *Delaware History* Vol. XXVII, Numbers 1–2, (Fall-Winter 1996–7), 4.

11. William L. Henson, 30. See also Don Timmons, "What the Delmarva Grower Survey Shows," *Broiler Business* (September 1976).

12. Robert H. Brown, "The Broiler Industry: the 'Sunday Dinner' Becomes Everyday Fare," *FEEDSTUFFS*, (29 October 1979), 98. See also Robert H. Brown, "Poultry Industry Integration: How it All Came About," *FEEDSTUFFS*, (29 October 1979), 103.

13. Bruce Drasal, telephone interview by author, Baltimore, Maryland. Mr. Drasal served as the Vice President of the United Commercial Food Workers (UCFW), Local 27 during the 1990s. The UFCW represented poultry workers on the Lower Maryland Eastern Shore.

14. Charles Laurent, "Is Integration Ahead?" *Poultry Digest*, Volume 14, No. 158 (April 1955), 193–196.

15. George W. Chaloupka, "The Early Days of Our Local Poultry Industry," Unpublished paper from his Private Collections (in author's possession).

16. William L. Henson, 30.

17. Mitzi Perdue, 43.

18. Ibid, 132.

19. Ibid, 27. All poultry growers were not on contract. There were independent growers and they made their own arrangements to sell their broilers to whomever they chose. Contract growers received their chicks from the big companies and signed a contract to follow the companies' specifications about raising and caring for the chicks until they reached the broiler (consumable) stage.

20. George W. Chaloupka, "The Early Days of Our Local Poultry Industry."

21. George B. Rogers, "Credit in the Poultry Industry," *Journal of Farm Economics*, XLV, 2 (May 1963), 409–415.

22. See the following, R. O. Bausman, "An Economic Study of the Broiler Industry in Delaware," University of Delaware Agricultural Experiment Station, Newark Delaware, Bulletin No. 242, (March 1943), 12 and Hugh A. Johnson, "The Broiler Industry in Delaware," University of Delaware Agricultural Experiment Station, Newark, Delaware, Bulletin No. 250 (1944), 5–57. Although this study focused on Delaware, poultry contractors there received the same contracts as growers on the Maryland side of the Delmarva because the majority of the growers raised chickens for the same large poultry companies.

23. See the following works, Paul Aho, "Broiler Grower Contracts in the United States," *Broiler Industry*, (October 1988), 26–31; "Contracts Will Change Fundamentals of Agriculture," *Feedstuffs* [Opinion Editorial], (18 April 1994), 8; Charles R. Knoeber and Walter N. Thurman, "Don't Count Your Chickens . . . Risk and Risk Shifting in the Broiler Industry," *American Journal of Agricultural Economics*, Volume 77 (August 1995), 486–496; Floyd A. Lasley, *The U.S. Poultry Industry: Changing Economics and Structure*, Agricultural Economic Report No. 502 (1983). Washington, D.C.: U.S. Department of Agriculture, Economics Research Service; Floyd A. Lasley, Harold Jones, Jr., Edward Easterling, and Lee Christensen, *The U.S. Broiler Industry*, Agricultural Economic Report No. 591 (1988). Washington, D.C.: U.S. Department of Agriculture, Economic Research Service; Louise Lee, "Weak Poultry Sales Are Putting Squeeze on Smaller Growers," [National Poultry Growers Association] Available at http://www.web-span.com/pga; Carole Morison, "Are the Independents Making a Come Back?" [National Contract Poultry Growers Association—1996] Available at http://www.web-span.com/pga/op_ed/mdag.html; Carole Morison, "Contract Poultry Farming," [National Contract Poultry Growers Association—1996] Available at http://www.web-span.com/pga/contracts/contfarm.html; "Poultry Processors Drive Growers to the Brink," [Progressive Populist] *Monthly Journal of the American Way*, (February 1996), and T. Vukina and W. E. Foster, "Efficiency Gains in Broiler Production Through Contract Parameter Fine Tuning," *Poultry Science*, Volume 75 (November 1996), 1351–1358.

24. Edward Covell, interview by author, Georgetown, Delaware. See also, "Perdue Offers You the Opportunity to start your own business: Grow with Perdue, There's never been a better time." *Perdue Farms* [Promotional Booklet], 1982. While this assertion is not explicitly stated in the promotional booklet, however, it is a reasonable inference that Perdue seemed to have targeted white growers as can be seen in their growers' ranks as well as their reduced credit risk.

25. Broilers are chickens raised for consumption. They are usually raised and kept for about nine weeks after which they are taken to the processing plants for slaughter as meat.

26. Raymond P. Smith, interview by author, Berlin, Maryland.

27. Edward Covell interview by author.

28. Mitzi Perdue, 28.

29. William H. Williams, *Delmarva's Chicken Industry. . .*, 51–54.

30. Ibid.

31. Roger Horowitz, 4.
32. Mitzi Perdue, 70.
33. Ibid, 115.
34. Ibid.
35. Charles C. "Chick" Allen, III, "Allen Family Foods Inc.," *Salisbury Daily Times*, 1 January 1970.
36. George W. Chaloupka, interview by author, Georgetown, Delaware.
37. Delmarvalous was an endearing term used by locals to promote a distinct brand of chickens produced on the Lower Maryland Eastern Shore. The term elicited pride in a locally inspired industry that rose to become one of the most successful and wealthiest companies in the nation.
38. Ed Covell, interview by author.
39. Robert J. Brugger, *Maryland: A Middle Temperament, 1634–1980* (Baltimore: Johns Hopkins University Press, 1988), 578–9.
40. Ibid.
41. John L. Skinner (ed.), *American Poultry History, 1823–1973* (Madison, Wisconsin: American Printing and Publishing, Inc., 1974), 176.
42. Gerald B. Truitt Jr., interview by author, Georgetown, Delaware. Mr. Truitt was a local banker on the lower Maryland Eastern Shore in the 1950s and 1960s.
43. Frank Perdue, interview by Julie Sloane, *Fortune* Magazine, http://www.fortune.com/smallbusiness. For example, between 1960 and 1970, the value of commercial poultry production was $110 million with the bulk of the poultry produced on the Lower Maryland Eastern Shore counties. See *Maryland Agricultural Statistics*, Annual Summary for 1970 [Poultry Statistics], 15. See also, Delmarva Poultry Industry, Inc., *Facts About Delmarva Poultry*.
44. Alison Morton, interview by author, Salisbury, Maryland.
45. See [Business Timeline]. Available at http://www.perdue.com/company/history/timeline; National demand studies conducted in the 1970s showed that both monthly and quarterly demands for poultry were elastic. See the unpublished research of Gerald R. Rector and Theresa Y. Sun on broiler demand analysis. Washington, D.C.: United States Department of Agriculture, Economic Research Service [n.d.].
46. "An Introduction to Perdue Farms" [Brochure] See also, http://www.perdue.com/history/timeline.
47. See Charles D. Schewe and Rueben M. Smith, *Marketing: Concepts and Applications*, second edition (New York: McGraw-Hill Book Company, 1980), 10.
48. See Maryland Department of Agriculture, *Maryland Agricultural Statistics*, Annual Summaries, 1971–1979.
49. Mitzi Perdue, 73.
50. This author made repeated visits to Perdue Farms' corporate headquarters in Salisbury and requested information concerning the number of workers. There were email requests for specific company-related information made as well, all of which were denied. However, the Delmarva Poultry Industry Inc., the trade association for commercial poultry producers in the region, estimated that by the 1980s, the total poultry personnel was 27,062 people (DPI Archives—*Facts About Delmarva's Broiler Industry, 1980 & 1985*). It should be noted here that there were no state lines

with regard to commercial poultry production on the Lower Shore because of the interstate nature of their operations. That is, Perdue operated in all three states and the numbers of poultry workers were not necessarily confined to those who worked in the three specific counties examined in this study but included those who worked in Perdue's plants in Delaware and Virginia.

51. See J. Frank Gordy, *A Solid Foundation: The Life and Times of Arthur W. Perdue* (Salisbury, Maryland: Perdue Incorporated, 1976), and Mitzi Perdue, *Frank Perdue. . . .*

52. Maryland Department of Agriculture, *Maryland Agricultural Statistics*, Annual Summaries, 1981–1989.

53. Carole Morison, interview by author, Pocomoke City, Maryland.

54. Gerald B. Truitt Jr., interview by author.

55. Perdue Farms, [Promotional Brochure] 1982.

56. "Favored" is used here not to suggest that the growers were wealthy. Instead, within the commercial poultry rank and file, growers occupied a relatively high status in the community. However, such high status did not negate their debt obligations. On the other hand and just as in other business ventures, poultry growers reaped great rewards if their flock performed well in terms of minimal diseases and other infections as well as a low mortality rate. In the local sense, a contract grower was a creative risk-taker. That is, he or she understood the risks involved and yet made the decision to creatively manage the risks.

57. This researcher could not locate information about the status of black contract growers during the 1970s and 1980s. Local interviewees corroborated that most of the growers were white. Repeated attempts to obtain information from the company were futile.

58. Carole Morison, interview by author.

59. The "Big Chicken," a label coined by critics of the industry, has come to signify big poultry companies—Mountaire, Perdue, and Tyson and their quest to produce meatier chickens in large volumes at a high cost to both people and the environment. As well, the label also related to an unequal and exploitative relationship between workers and management. To be sure, workers did not typically use this phrase. Rather, it was used mainly by critics of big commercial poultry companies.

60. Carole Morison quoted in Gretchen Parker, "She gives poultry workers a voice," http://www.delawareonline.com (6 June 2004).

61. See the following: James Bock and Dail Willis, "Changing face of the Shore; Latinos: Lured by abundant jobs and good wages, Hispanic immigrants have become a sizable community on the Delmarva Peninsular," *Baltimore Sun*, 13 October 1996; Rick Thornton, "Poultry growers feel trapped in power struggle," *Daily Times*, 25 March 1992; Christopher Thorne, "Farmers crying foul over Perdue," *Las Vegas Review Journal*, 25 December 1999, and Deborah Thompson Eisenberg, "The Feudal Lord in the Kingdom of Chicken: Contracting and Worker Exploitation by the Poultry Industry," *The Public Justice Center*, [n. d.].

62. I am using "dependent" here because the contract growers could not produce without the companies supplying them with flock. It was to the companies' advantage that they dictated when and how much flock the growers were to receive. On the other

hand, it fostered a dependency cycle that did not produce beneficial results for growers. These growers' contracts could be discontinued or cancelled at any time without notice. Also Edward Covell, interview by author, Georgetown, Delaware.

63. In fact, the 1990s was a busy decade for Perdue as it expanded into the Midwest and Southeastern United States as well as into more than 50 foreign countries. For example, it purchased several small to mid-size operations in Georgia and Tennessee.

64. Available at http://www.perdue.com/company/news.

65. See Maryland Department of Agriculture, Maryland Agricultural Statistics, Annual Summaries, 1990–1999.

66. Tyson Foods Inc., Annual Report, 1997, p.6. Perdue and the other companies' annual reports were unavailable to the public. This researcher made multiple attempts to obtain these annual reports from the Perdue, Allen, and Mountaire to no avail. Tyson's Annual Report was available only in accordance with the Securities Exchange Act of 1934 that required the filing of annual reports by publicly traded companies, Tyson Foods Inc., registered and incorporated in Arkansas in 1947 and reincorporated in Delaware in 1986. However, Tyson had a commercial poultry processing plant on the Lower Maryland Eastern Shore.

67. This author made an attempt to obtain an actual number of black workers in the commercial poultry production but was unsuccessful. The companies do not share such information with the public. However, some of the interviewees confirmed that the number of African Americans who worked in the processing plants increased after commercial poultry production became large-scale.

68. "Industrialization" is used restrictively in the context that commercial activities of any sort other than farming that brought diversity in local lifestyle qualified as industrialization. While the word clearly connotes factories and the like, its usage here makes a distinction between a lifestyle that was almost entirely dependent on farming and one that included leisurely activities as in the urbanized areas.

69. Spotwood Jackson, telephone interview by author, Baltimore, Maryland. Mr. Jackson has worked in the commercial poultry industry for more than 20 years and was one of a handful of African American line supervisors at the Salisbury poultry processing plant.

70. See Solomon Iyobosa Omo-Osagie II, "Count Her In . . ." *Southern Historian: A Journal of Southern History* Vol. XXIV, (Spring 2003), 40–49.

71. Enez Stafford Grubb, interview by author, Cambridge, Maryland.

72. Samuel J. Deal, an African American Eastern Shore native who attended the integrated James M. Bennett High School in Salisbury in the 1960s, stated that there were black high schools in Somerset and Worcester counties in 1950 and that there may have been a handful of blacks who completed high school in Somerset and Worcester counties. He contended that the questions about completing high school may have been posed in different ways for blacks than for whites, leading to the possibility that some blacks may have misunderstood the question about completing high school. Another contention was that the record keeping for blacks in the three counties during this time period was not the best or as thorough and reliable as one might have expected.

73. The word, "protracted" is used in this context to make the point that some of the workers' children and family members have worked in the industry in some capacity in the last three decades. While the younger generations, those born in the post-1980s may no longer show an interest in becoming poultry workers, many families in these counties have been impacted by commercial poultry production.

74. Alison Morton, interview by author.

75. What I mean by reliable here is that despite the low wages for their labor, blacks could always be counted upon to show up for work. In a way, the workers had little choice but to show up for work as poultry labor was one of the easiest jobs to do in the sense that workers did not need extensive training to work on how to kill, de-feather, clean, and package the chickens on the conveyor belts. Also, the threat of Hispanic workers hung in the air if black workers did not show up; the Hispanic workers readily stepped in and performed the work that was needed.

76. Ziaul Z. Ahmed and Mark Seiling, 36. In subsequent years, studies have shown poultry workers as more likely than workers in other industries to sustain injury on the job. See also, "Injury and Injustice—America's Poultry Industry," http://www.ucfw.org.

77. The author made repeated attempts including telephone calls and actual visits to the corporate headquarters in Salisbury to obtain data but was unsuccessful. The company's spokesperson cited privacy concerns. However, a current plant supervisor, Spotwood Jackson, in a telephone interview corroborated the interviewees' assertion that the majority of the line workers inside the plant have been African Americans.

78. Hattie Beckwith, interview by author, Salisbury, Maryland. Ms. Beckwith was a community activist who fought for social justice in the Salisbury black community during the 1960s and 1970s.

79. "Official," as used in this sense, denotes pronounced exclusionary ordinances. Unofficial segregation denotes local customs and practices of people (blacks and whites) knowing their place and not stepping out of their boundaries. There were many unwritten ordinances that kept blacks from engaging in the lucrative tourism business in places like Ocean City prior to the 1980s to maintain an image that whites wrongly thought improved local tourism: a tourist haven for whites' relaxation.

80. For example, data from the U.S. Department of Labor's Bureau of Labor Statistics showed that in 1987, the average hourly wage for a poultry worker was $5.87.

Chapter Five

Commercial Poultry Production, Public Health, and the Environment

Debate has raged over the impact of industrial animal production on food supply, public health, and the environment. Much of the literature on commercial poultry production has shown an incontrovertible link to public health and the environment.[1] Although, consumption of animal-related meat product is an important source of essential nutrients needed for human growth, some of the diseases found in these animals became hazardous to public health. The various aspects of commercial poultry production and how they had both short and long-term implications on public health and environment of the Lower Eastern Shore is a reoccurring issue in commercial poultry with regard to health concern. The use of antibiotics to both treat poultry diseases in humans as well as to fatten chickens to reach a desired size and appearance has created further public health and environmental concerns.

The major environmental concern was that commercial poultry production had a polluting effect on the Chesapeake Bay because of the industrial wastewater runoffs that emptied into the Bay. The runoffs negatively affected maritime commercial agriculture, which served as a source of livelihood for many citizens of the Lower Eastern Shore. In effect, the industry's impact in this region extended into the health and environmental realms. Although the industry was a beneficial venture, it became a major issue of concern for public health practitioners and environmentalists who argued that poultry producers failed to invest in ways that reduced the public's exposure to poultry-related diseases. In essence, health and environmental issues informed the debate on the dichotomy of poultry as an economic livewire of the Lower Eastern Shore, and as a source of health and environmental hazards.

COMMERCIAL POULTRY AND PUBLIC HEALTH

There were persistent chicken-related diseases such as Newcastle Disease Virus (NDV), Marek's Disease, and Infectious Bronchitis that threatened the viability of the early phase of the commercialization of poultry production.[2] Diseases and viral outbreaks were common place during the early to the large-scale phases of commercial poultry production. In one sense, commercial poultry served a nutritional value, but on the other, it exposed society to diseases and food poisoning, which led to serious public health crisis. There were a variety of reasons for these diseases among which included poor management of flock, poor sanitation of the poultry houses, and improper breeding within the progenies.[3] The producers at the time had yet to develop effective ways to manage poultry waste. Some producers lost many of their flock because the micro organic contents of fecal wastes contained high levels of toxicity.[4]

The early period of commercial poultry on the Lower Eastern Shore was characterized by a paralyzing chicken egg disease that affected mostly independent and subsistent producers. In the 1930s and 1940s, poultry growers were responsible for maintaining the health of the chickens which were susceptible to respiratory diseases, the avian flu, epidemic tremors, and Coccidiosis.[5] In 1939, "Coccidiosis was the number one problem facing growing chicks" and it gave both poultry growers and producers great deal of concern.[6] By the 1940s, the discovery of sulfas by P. P. Levine of Cornell University proved pivotal for the commercial poultry industry.[7] On the Lower Eastern Shore, sulfas, including sulfamethazene, sulfaquinoxaline, and sulfaquanidine were used to treat coccidiosis in commercial flocks.[8] Prior to the use of sulfas, it was common for growers to sustain 25 percent mortality rate among their flocks but the application of sulfas greatly reduced the chicks' mortality rate.[9]

Public health was a major issue of concern in commercial poultry production since the 1930s because of the impact that chickens had on consumers. In the early phase of commercial poultry production, 1930 to 1950, different types of bacterial infections were known to be present in poultry meat. Experimental studies detected the presence of *Enterococcus* spp. in selected poultry production facilities on the Delmarva Peninsula, which was capable of infecting humans and could prove fatal if left untreated.[10] If treated with the wrong medication or an ineffective antibiotic, it could lead to the development of drug-resistant food poisoning bacteria.[11] The two commonly used antibiotics, in the feed for the flock, were Oxytetracycline and Chlortetracycline.[12] There was public concern about antibiotics because when used for the "wrong" reasons and on a more than usual basis, it became ineffective in the sense that it was unable to fight off bacteria.[13] The inability of the antibiotics to fight off poultry-born bacteria because of overuse affected public health, because as more people consumed poultry meat they became more exposed to bacteria that were antibiotic resistant.

Studies by Johns Hopkins University toxicologist Ellen K. Silbergeld and other researchers, which focused on commercial poultry production and the health of the workers on the Lower Maryland Eastern Shore, have shown that "the antibiotics used in chicken feed might be creating tough, drug-resistant bacteria that cling to workers' hands and wash off into rivers."[14] Further studies that have been conducted on the impact of industrial animal production in general and commercial poultry production in particular found that the feed given to industrial animals, like poultry, undermined the effectiveness of many antibiotics that were used to treat various diseases in humans as a number of bacteria became resistant to antibiotics.[15] Commercial poultry production on the Lower Maryland Eastern Shore harmed the environment and threatened public health because of "the release in high levels of gases, odors, nutrients, pathogens, and antibiotic resistant bacteria into the air, water, and soil."[16] The method of poultry waste disposal from the production facilities created more problems and challenges for the environment, in particular, ground water pollution and public health.[17]

Some of the bacteria found in chickens became resistant to antibiotics because of overuse.[18] The antibiotic by itself was not the public health issue, rather, the argument was that antibiotic use in poultry feed was having a direct effect on the development of resistant strains of bacteria.[19] The concern over the indiscriminate use of antibiotics did not stop commercial poultry producers from using such growth antibiotic.

The concern over the public health implications of commercial poultry increased with the emergence of large-scale commercialization of poultry production in the 1950s. There was a public safety and welfare interest on the part of the government to ensure that the public was protected from possible health threatening outbreaks of diseases. The contending issues at the onset of large-scale commercialization of poultry were selling poultry meat in an unwholesome and unadulterated form, as well as misbranding. Consumers were exposed to health hazards in the sense that some poultry producers did not follow adopted government safety standards such as labeling and branding their poultry to safeguard the public. As commercial poultry became an interstate commerce, it fell within the jurisdiction of the federal government in terms of laws and regulations to ensure the public's health. Although Maryland along with other states had the power to regulate their products, the interstate nature of commercial poultry production led to the enactment, by the U.S. House of Representatives, of the Poultry Products Inspection Act (PPIA) of 1957, which mandated, among other things that,

> Each official establishment slaughtering poultry or processing poultry products for commerce or otherwise subject to inspection under this chapter shall have such premises, facilities, and equipment, and be operated in accordance with such sanitary practices, as are required by regulations promulgated by the Secretary for the purpose of preventing the entry into or flow or movement in commerce or burdensome effect upon commerce, of poultry products which are adulterated.[20]

By the 1950s, commercial poultry's profitability gave way to poultry market manipulators who attempted to make profits at the expense of public health. Slaughterhouses emerged, which did not meet safety and sanitary standards.[21] The end result was the presence of infected and contaminated meat on the market. The federal government intervened to check against such contaminations and infections. The enactment of the PPIA was timely because contamination and infections occurred most frequently at the processing phase due to exposure to multiple microorganisms. The act effectively brought commercial poultry under federal regulation and mandated that all slaughtered and processed meat must be inspected before they were marketed for consumption. This mandated inspection was aimed at minimizing the chances of contamination and infection, which increased with large-scale production. *Salmonella*, along with other types of contaminants such as Yesinia and Listeria were found to be present in commercial poultry waste, chickens and other poultry related products and caused food-related poisoning.[22] Some studies have shown a correlation between *Salmonella*, *Campylobacter*, and public health.[23]

Commercial poultry workers faced numerous health risks, particularly those that worked in the chicken houses who suffered from multiple health complications, especially respiratory problems. There were numerous documented cases of respiratory illnesses, which workers contracted from "breathing in chicken litters and ammonia fumes generated by manure in confinement poultry houses."[24] Respiratory diseases were prevalent among poultry workers on the Lower Maryland Eastern Shore. But since African Americans dominated the poultry working class on the Lower Eastern Shore, it was likely that respiratory diseases were more prevalent among them than their white counterparts.[25]

Many workers were exposed to respiratory illness because they were not supplied with masks and other equipment that protected them from exposure. Incidentally, respiratory illness was not only confined to poultry workers on the Lower Maryland Eastern Shore. There was a study on exposure to dust-borne bacteria in poultry houses and its effect on workers, which was conducted in Poland from 1966 to 1975. This study showed that the exposure and presence of dust created a serious problem to occupational agriculture and by extension those who worked in commercial poultry production.[26] Due to the confined space in which chickens and humans interacted, both the workers and the chickens were vulnerable to mutual infection with air-borne disease agents such as Newcastle Disease Virus.[27]

Also in a 1984 study, Salisbury State College respiratory therapy instructor, Xavier Stewart, found a correlation between dust and lung disease. The study demonstrated that "dust," such as that found in the chicken houses, "caused problems indicative of lung disease."[28] The study showed a correlation between public health and commercial poultry production. A consistent exposure to the poultry litter and its dried fecal content emitted toxins that affected the respiratory tracks of the workers. Despite the outcome of this study, it seemed that the companies invested more resources in the chickens' health than they did on the health of the workers. When the workers became ill, it was considered a routine hazard that came with every job.[29]

In the chicken houses, as the catchers waded through hordes of chickens and chicken waste, they contended with dead chickens and those that were near death. These decaying chickens carried disease agents that were harmful to humans upon contact with the skin or even when the air was inhaled. There was a concern about the conditions inside the poultry houses as poultry researchers worked in the 1980s to explore ways to better dispose of dead chickens. The research was not to determine how best to change the environment and minimize the catchers' exposure to harmful disease agents. Rather, the focus of the research was more on how to help make the chickens healthier while producing them at a lower cost.[30]

Researchers developed preventive ways to minimize contamination in efforts to improve sanitary standards. Chlorine, which contained agents capable of killing microbes, was used to sanitize the water in which the chickens were dipped. Sanitizing the water with chlorine posed public health danger because of high amounts of chlorine.[31] This practice was commonly used in the processing plants on the Lower Maryland Shore just as it was used in other commercial poultry producing areas. The inhalation of chlorine, which was very toxic, affected the sensory organs of the workers. While it was understandable that the chickens had to be healthy and free of diseases, the workers' health should also have been of equal concern. Both the chickens and the workers needed to be healthy while ensuring minimum mortality of the flock.[32]

To maintain the momentum for consumption and demand, throughout the periods of commercial poultry production, nutrition for the chickens was a considered priority. Producers used alfalfa meal, distillers' grains, dried whey and wheat midds as part of the poultry diets during the 1950s.[33] These mixes ensured that the chickens were healthy and that the producers were assured of some measure of profit.

The result of these diets varied when fed to either male or female chickens. It also varied depending on the particular strain. As consumers preferred meatier chickens starting with Perdue's marketing of meatier chickens in the 1970s, producers focused on the chicken strains such as Arbor Acre that grew faster and therefore produced higher gains for commercial poultry producers. Although they yielded better result than other strains, Arbor Acre had a higher mortality rate. The medicated feeding of the commercial flock, posed public health concerns in commercial poultry production.[34] Arbor Acre may have yielded better results in terms of how fast they reached consumable stage, but they required a more medically concentrated feeding to grow faster than other strains such as Leghorn or Rhode Island Red Cockerel. The consumption of this brand of chicken posed a public health concern because of long-term health effects on humans resulting from the medicated feeding.

In the 1980s, there were documented cases of poultry-related contaminations that impacted public health. A Congressional report authorized by the subcommittee on intergovernmental relations and human resources of the U.S. House of Representatives Committee on Government Operations found that the U.S. Food and Drug Administration (FDA) failed to "monitor the use of toxic drugs and nutrition supplements in raising livestock, thereby posing great threat to the health of consumers."[35] This finding reinforced researchers' concern over the impact of commercial poultry production on public health.

The decade of the 1980s featured stiff competition among commercial poultry producing firms as each company sought to convince consumers that their brand of chickens were healthier, and more attractive. A key reason that many people consumed poultry meat was because of its low-fat content. Perdue Farms invested heavily in efforts to produce low-fat chickens that gave them a competitive advantage over other competitors. The company spent $2.5 million in 1985 on the production of low-fat chickens.[36] Perdue Farms was in a better position to make such an investment because by the 1980s, the company had risen to being among the top ten firms in commercial poultry production and was ranked the third largest commercial poultry producer in the nation.[37]

One major issue of concern about public health in the 1980s was the lack of strict inspection of poultry facilities and processing methods. The standard method of inspection through the 1980s was visual inspection. The meat and poultry inspectors walked through the assembly lines of the processing plants and visually inspected the chickens for any signs of disease or deformity as they came through the conveyor machines. Afterwards, the inspectors pronounced the chickens safe, especially if there were no signs of bruises on the chickens. The problem was that even if the chickens were contaminated, it was difficult to detect this through mere visual inspection.[38]

The public health concern was that the invisible contaminants in the chickens were overlooked and subsequently the bacteria were then passed on to the consumers.[39] This concern led to the replacement of visual inspections in the late 1980s, with another method called spot-checking. Under this new method, the inspectors had to physically examine each chicken as it came through the processing line. This way, they had a higher probability to detect bacteria such as *salmonella* or *campylobacter* in the chickens.[40] This new method was effective because it gave the inspectors a better view of the chickens and an ability to isolate contaminated and infected chickens to keep them from contaminating and infecting others in the critical processing phase.

During the 1990s, there were issues of public health and poultry that persisted. *Salmonella* and *campylobacter* remained as the major causes of foodborne illness such as gastrointestinal illness among the consuming public. The challenge for both poultry producers and the government agencies such as the FDA and USDA was how to address these bacteria-related contaminations. The aim was to reduce or minimize the frequency of the infections or eliminate them from the food chain. In 1995, and in an effort to control outbreaks of e-coli infections in the poultry houses and reduce their transmission to humans, the FDA approved the use of sarafloxacin, an antibiotic that was mixed with chicken drinking water. This drug was intended to reduce the spread of e-coli as well as help poultry farmers reduce the rate of chicken mortality that resulted from non-treatment of e-coli infections.[41]

Cases of food contaminations were reported in New York and the mid-west during the 1990s and scientists and researchers sought ways to help contain the spread of the contaminations. One proposed solution to address these contaminations was irradiation. The process of irradiation called for each chicken in the processing plant to "be placed on a conveyor belt, which then travels into a chamber protected by thick concrete walls. A radioactive source such as cobalt 60 or cesium 137, a waste product of the nuclear power industry, emits the radioactive rays,"[42] that were supposed to kill the microorganisms found in poultry meat. Commercial poultry producers opposed the irradiation of their poultry "because of strong consumer resistance to irradiated foods."[43]

This method of controlling food contaminants was so controversial that even researchers and scientists posited divergent arguments about irradiation and the effects on public health.[44] The opponents of irradiation were not entirely convinced about its effectiveness and were concerned about its possible long-term adverse impact on consumers. Poultry producers on the Lower

Maryland Eastern Shore did not actively engage in this debate because the practice was not widely used among them. Although this was the case, some companies initiated their own measures to control food contamination. For example, Perdue Farms developed self-imposed, as well as industry-wide health and safety standards. It built food safety processes such as Microbial Reduction plan that enabled early detection of *salmonella*, *campylobacter*, and other disease-causing bacteria into its commercial poultry production.

One solution for addressing *salmonella* contamination was to dip the chickens in a chemical solution, trisodium phosphate, which was used in the processing plants, to kill the bacteria before they were finally packaged and shipped out for consumption.[45] The USDA supported this method of *salmonella* treatment after the FDA found it safe in terms of public health considerations.[46] Another technique was invented by a collection of USDA and private sector scientists towards the end of the 1990s. These scientists introduced "CF-3 or Preempt, a mixture of beneficial microbes that occur naturally in chicken."[47] Despite the various techniques proposed to treat these bacterial infections in commercial poultry production, the concern about their impact on public health persisted.

The significant public health issues regarding commercial poultry production during the 1990s, remained food contamination and how to prevent and reduce its occurrence. In addition to the earlier proposals such as irradiation and chemically dipping the chickens in solutions, a series of executive actions were taken in an attempt to allay public concern over the debate on irradiation and chemical dipping. In 1993, President William Jefferson Clinton ordered the hiring of 160 meat and poultry inspectors to ensure the enforcement of federal inspection standards.[48] Since commercial poultry production was an interstate commerce, companies on the Lower Maryland Eastern Shore, including Perdue Farms, were affected by the president's order to enforce inspection standards. The order was meant to restore consumer confidence in poultry meat.

The president ordered for more meat inspectors because there was a shortage of inspectors resulting from the effects of the President Ronald Reagan era of deregulations and cuts in social and environmental programs of the 1980s. Cuts in the budgets of the Health and Human Services Department and USDA, parent agencies of FDA and the Food Safety and Inspection Service (FSIS), the agency that oversaw compliance of food processing plants also contributed to the shortage of inspectors.[49] These inspectors ensured that commercial poultry plants were in compliance of federal regulations on poultry slaughterhouses.

In the late 1990s, the FDA allowed food producers, including commercial poultry producers, the flexibility of deciding whether to, or not to, label their foods as irradiated.[50] While Perdue and other commercial poultry producers sought measures to inspire public confidence in poultry consumption, the reality was that animal disease always lingered. As humans consumed poultry, the threat to public health remained. Researchers and public health practitioners helped reduce the frequency of diseases in commercial poultry production.

COMMERCIAL POULTRY AND THE ENVIRONMENT

The commercial poultry industry was a profitable venture for the local economy of the Lower Maryland Eastern Shore. There were environmental issues such as air and water pollution, and land degradation that arose. In fact, commercial poultry production has been described as "a major cause of environmental degradation in the United States" because of the multiple adverse effects the industry had on the environment such as the detrimental effects on fish and other wildlife.[51]

Historically, the environment on the Lower Eastern Shore has been a focus of attention in terms of the adverse impact of commercial poultry production in part, because of the Chesapeake Bay and its maritime importance.[52] Activities on and around the Bay have had broad implications on the environment and public health. The disposition of wastes from commercial poultry production facilities carried contaminants, which affected the Bay.[53] These wastes carried toxins that led to diseases and outbreaks of diseases, which were caused by certain harmful bacteria present in poultry waste. When fishes made contact with the wastes, it led to outbreak of other bacteria-borne illnesses that sickened consumers of maritime-related products.[54]

Commercial poultry production flourished on the Lower Maryland Eastern Shore and it increased the volume of waste production that in turn led to an increased volume of environmental pollution since chickens produced daily wastes. The (NIMBY)[55] "Not in my back yard" syndrome was an issue in determining where to dispose this poultry waste. It was an issue because independent and contract poultry producers as well as large-scale commercial poultry companies were saddled with the burden of disposing the large volumes of wastes produced.

The concern for environmental damage, which centered on the Chesapeake Bay,[56] was the fact that commercial poultry producers, in an attempt to reduce costs of transportation, incineration, and composting of excess poultry manure, dumped these wastes in open space and sometimes in larger quantities than the lands were able to handle.[57] Due to its geography, much of the streams, creeks, or estuaries on the Lower Eastern Shore converged onto the Bay, and all activities such as dumping excess manure on the open land ran off either to the Bay, or to one of the many tributaries that emptied their contents into the Bay.

The arable lands of the Lower Eastern Shore made the area very conducive for crop farming such as fruits and vegetables. Its proximity to sources of water for farmlands gave the region an appeal that made farming an attractive occupation other than fishing, crabbing, or oystering. Manure was therefore of great importance to the farmers, but much of the manure generated by the poultry industry was unusable as fertilizer because of its phosphorus and nutrient contents. The manure caused more damage for land farmers such as soil degradation. In addition, the resulting runoff from the processing plants impacted the health of the Bay, which "deteriorated largely as a result of nutrient overenrichment, concomitant reduction in light availability, and loss of habitats that provide complexity."[58] The runoffs also contained hazardous metals such as Fe, Zm, Cu, and Cd in general heavy metals and compounds that increased human risks through fish consumption.[59]

However, managing these poultry wastes and manure was problematic.[60] According to the Maryland Department of the Environment, poultry manure made up about "40 percent of the nitrogen and 48 percent of the phosphorous entering into the Chesapeake Bay from Maryland's Eastern Shore."[61] Different methods of disposing excessive manure and waste were proposed. One of these methods included burning poultry manure to generate energy.[62]

The Lower Eastern Shore was the epicenter of commercial poultry production in Maryland, and it generated a high percentage of the 355,000 tons of poultry litter a year that contained high levels of nitrogen and phosphorous.[63] The more poultry produced, the more the space needed to dispose poultry waste. During the early period of commercial poultry production, there were no formalized ways of disposing poultry wastes.[64] In essence, poultry waste disposal during the 1930s was the prerogative of the producers. The producers simply took their wastes to the community landfill. The issue of waste disposal was a major source of tension between commercial poultry producers and environmentalists. In an attempt to manage waste, many suggestions were proffered about the use of poultry waste that addressed both public and environmental health concerns.[65]

Each of the commercial poultry companies on the Lower Eastern Shore such as Mountaire Farms, Tyson Foods, and Perdue Farms dealt with environmental concerns in the sense that each company generated large amounts of poultry waste that needed to be disposed. Perdue Farms was the most affected of the three companies with regard to poultry waste management, because it had more plants in Maryland than Mountaire Farms or Tyson Foods. As a result, Perdue Farms shouldered most of the blame for environmental pollution. Perdue Farms portrayed itself as an environmental steward and declared that,

> . . . it [was] committed to environmental stewardship and shares that commitment with our farm-family partners. We believe it is possible to preserve the family farm and provide safe, abundant, and affordable food supply while protecting our communities and the environment. Perdue farms leads the industry in addressing the full range of environmental challenges related to animal agriculture and food processing, investing millions of dollars in research, new technology, equipment upgrades and awareness and education.[66]

While Perdue Farms' statement may have seemed altruistic, it appeared that the company was forced to embrace environmental preservation after much criticism from environmental groups. Had the company been truly committed to protecting the environment, perhaps it would have shown its commitment on its own by reducing its environmental pollution. The resources that Perdue Farms needed to commit to environmental preservation would have led to major financial loses in terms of the costs that were involved with disposing the waste generated in commercial poultry production. Implicit in Perdue's declaration of commitment to environmental stewardship was the recognition of the impact that commercial poultry production had on the environment.[67] Large-scale poultry production required millions of gallons of water used in the processing plants to clean chickens on a daily basis. The risks became greater and the concern of environmentalists was that the water from the plants contained deadly pathogens which wounded up either on the lands where feed grains were cultivated, or back into the unpolluted natural waters of some tributaries that emptied into the Chesapeake Bay.[68] The pathogens in the waters endangered commercial seafood activity such as oysters and crabs.

The Congress responded to the environmental concerns by enacting the Comprehensive Environmental Response, Compensation, and Liability Act (CERCLA) in 1980 after scientific studies showed the threat that commercial poultry production posed to public health and the environment.[69] This act regulated hazardous substances including those from commercial poultry production plants such as gaseous ammonia that were present in poultry operations, which endangered public health and the environment. Ammonia was used in the farm as a fertilizer and was present in poultry litter.[70] Prior to the enactment of this act, it was routine for farmers to openly dispose of the accumulated ammonia due to large poultry wastes. This act was intended to end this practice of environmental pollution.

In the 1980s, environmental protection and preservation had become rallying points for communities across the nation. On the Lower Maryland Eastern Shore, the focus continued to be on the Chesapeake Bay in part because of its extensive implications in terms of maritime commerce. A series of annual reports linked commercial poultry production to environmental degradation.[71] These reports became the catalysts for additional governmental intervention in environmental protection.[72] Congress enacted the Emergency Planning and Community Right to Know Act of 1986 (EPCRA).[73] This act was an empowerment tool for communities to demand that they be informed and made aware of chemical emergencies in the event of accidents at companies operating in their communities. The act also held industrial companies including commercial poultry producers accountable for informing the communities in which they were located, in the event of company actions and activities that affected public health and environment of the communities. Prior to this act, industrial companies such as commercial poultry producers, were not obligated to inform their host and surrounding communities on environmental matters that may have affected public health, polluted the environment, and wellbeing of such communities.

In 1990, Congress sought additional measures to control the pollution from animal industrial production by passing the Pollution Prevention Act (PPA). The main focus of this act was on finding ways to reduce pollution by deploying practices that improved efficient use of energy, water, and related resources. The act was also intended to function as a preventive measure to conserve as well as reduce ways to prevent pollution from taking place.[74] In a sense, this preventive legislation benefited even commercial poultry producers because energy conservation reduced overhead costs. One notable provision of the EPCRA and PPA was the requirement of industrial and commercial companies to file a yearly Toxic Release Inventory (TRI) with the Environmental Protection Agency (EPA). They were required to file this report because they used certain harmful toxic chemicals in their production facilities that were detrimental to community health. This report mandated the affected companies to reveal the amount of toxic chemicals that were released into the air, injected into the ground, or disposed into the waterways. Perdue Farms reported that between 1985 and 1995, it used quantities of ammonia, sulfuric acid, and sodium hydroxide in its plants on the Lower Maryland Eastern Shore as manufacturing and ancillary aids, but that there were "no non-zero releases or transfers."[75]

The reality of commercial poultry production was that manure and run off from the poultry processing plants had to be disposed. Even then, for cost-effective changes in the production of poultry to take place, it meant that the producers had to develop innovative ways and minimize their costs of producing poultry. A shift in the production methodology was not as much of a financial burden for the big companies as it was for small or independent farmers, except in a few instances where the companies helped the farmers disposed their accumulated waste.[76]

During the 1990s, commercial poultry production was linked to the fish-killing *Pfiesteria piscicida* outbreak, which invaded the Pocomoke River and the surrounding waters, killing millions of fish. Scientific and environmental studies linked the outbreak to multiple pollutants contained in runoffs from commercial poultry processing plants.[77] However, the link was hotly debated by commercial poultry farmers, fishermen, and others engaged in maritime agricultural commerce.[78] In addition, the fishing community showed neurological symptoms from the toxins produced by large overgrowths of *Pfiesteria*.[79]

For the local communities on the Lower Shore, poultry litter, believed to be connected to *Pfiesteria*, posed environmental hazards to their water supply. In 1997, Perdue Farms, Inc. was sued by the Maryland Department of the Environment, which charged that one of Perdue's numerous processing plants dumped organic waste into the local Wicomico River contaminating another source of livelihood for the local residents: fishing. Perdue agreed to pay $380,000 in penalties and $150,000 in remediation.[80]

Allegations of environmental pollution against Perdue and other poultry companies were widespread during the 1990s. Due to its large volume of commercial poultry production, its poultry waste, known to harm water quality, ran off into the local streams and other tributaries.[81] Poultry and the Maryland Lower Shore were intertwined, but there were some disagreements between the communities and the industry over the negative impact of poultry production on the environment. Poultry companies were accused of violations of environmental laws.

In 1997, the U.S. Department of Justice alleged that Hudson Company, a subsidiary of Tyson Foods Inc. on the Lower Maryland Eastern Shore violated the Clean Water Act in its operation of Tyson's [Hudson] Berlin commercial poultry processing plant by failing to disclose its violation of the National Pollutant Discharge Elimination System (NPDES). Hudson, along with Tyson entered into a Consent Decree agreement with the Department of Justice to pay $4 million and another $2 million towards environmental projects.[82] Hudson Food had a history of public health and environmental violations. In 1997, the Occupational Safety and Health Administration (OSHA) found that the company had a persistent pattern of workplace safety and worker violations in its poultry processing plants such as the one located in Noel, Missouri.[83]

In 1997, Perdue Farms was sued by the State of Maryland for exceeding the required limits of allowable hazardous byproducts, which traveled into a local creek in Worcester County.[84] The following year, the Maryland Department of the Environment filed a criminal complaint against Tyson Foods in the Circuit Court of Worcester County, and alleged that the company violated Maryland's water pollution control laws at its Berlin plant. The company denied the charges but later paid fines for environmental pollution.[85] Also, "in 1998 Tyson Foods was fined $6 million for pollution of the Kitts Branch waterway in Worcester County, which had become loaded with coliform bacteria, phosphorous, and nitrogen dumped by a single poultry slaughter plant in Berlin."[86]

These cases were emblematic of the delicate balance between commercial poultry production on the Lower Maryland Eastern Shore and the local residents' attempts to preserve their environment on one hand, and the need to sustain the local economy on the other. The Tyson plant was important to the local Shore economy because at the height of its operation, the Berlin processing plant employed about 500 workers.[87]

ENVIRONMENTAL ACTIVISM AND COMMERCIAL POULTRY

Activism against the negative environmental effects of commercial poultry production was given much impetus by the activities of the Chesapeake Bay Foundation (CBF), which was formed in 1967. The mission of the CBF was to ". . . restore and sustain the Bay's ecosystem by substantially improving the water quality and productivity of the watershed and to maintain a high quality of life for the people of the Chesapeake Bay."[88] The CBF carried out an aggressive agenda of environmental protectionism and restoration. It educated the public about the valuable assets of the Bay. The CBF remained in the forefront of advocating for a healthy environment. Due to its primary focus on protecting the Bay, the CBF was one of the strongest and most persistent critics of the commercial poultry industry. It charged among other things, that the poultry industry's production methods and the methods of disposal of its wastes polluted the environment and caused health hazards to the consuming public.[89]

In the late 1990s, the Bay Friendly Chicken (BFC), made up of chicken catchers, workers, marketers, academics, environmentalists, and clergy was organized on the Lower Eastern Shore. The group proposed the feasibility of establishing the BFC as a community stock-held poultry company. The plan called for BFC to produce affordable chickens with social and environmental considerations anchored on seven main principles.

First, they argued for a local control of such a commercial poultry producer; second, they proposed a decision making process with the stakeholders as the major players; third, they wanted high quality birds that were free of antibiotics, hormones, and other medications; fourth, fair treatment of commercial poultry growers who they argued had been exploited by integrated companies such as Perdue and Tyson; fifth, high environmental standards including employing technologies and systems that protected environmental pollution; sixth, high labor standards and livable wages and adequate benefits for workers, and seventh, formalize the aforementioned commitments in rigid bylaws that met corporate compliance.[90] However, the most significant reason for the proposal of this enterprise was to minimize the environmental hazards posed by commercial poultry producers on the Lower Maryland Eastern Shore. The persistent pollution of the Bay through run offs from commercial poultry processing plants as well as the manure and waste that sipped through the soil and emptied into the Bay were compelling reasons for the proponents of the BFC. Also, the implication of toxic contaminants from exposure to commercial poultry plants to public health was another motivating factor for the proposition of the BFC.

The BFC became commercial poultry's fiercest critic and argued for an alternative organic chicken with minimal damage to the environment and the Bay. The group's argument was informed by the fact that as the largest estuary in the United States, the Bay was a valued treasure in Maryland.[91] Therefore, it was in the best interest of the proposed stakeholders to support the BFC concept. While the idea of an organically produced poultry meat seemed to have resonated, commercial poultry production was firmly entrenched on the Lower Maryland Eastern Shore. The well established commercial poultry producers like Perdue, Tyson, and Mountaire had begun growing organically produced chickens as part of their marketing strategy to meet consumer varying demands. This strategy countered the potential competition that would have been posed by solely organic commercial poultry producers such as was proposed by the BFC. The BFC concept was an alternative to the medically-dependent commercial poultry, and that an organic and non-medicated poultry could meet the dietary needs of consumers while friendly to the Bay.[92]

Another major contention by environmentalists was the millions of gallons of water that were expended every day in the processing plants for cleaning chickens before they were shipped out to super-market shelves. Previous studies had concluded among other things that "iron reducing and sulfur forming microorganisms may be the major clogging agents in these soils flooded with poultry processing wastewater."[93] These environmental concerns persisted through the 1990s as commercial poultry production on the Lower Maryland Eastern Shore expanded. The concern arose from the fact that commercial poultry producers in Wicomico and Worcester Counties used about 3 out of the nearly 5 million gallons of water used daily at the processing plants.[94]

Environmentalists also pointed out that the water from the treatment plants found their way back to the Bay and other tributaries causing the contamination of sea-borne foods such as fish, oysters, crabs, and shrimps. This contamination exposed the public to health risks, learning and memory difficulties.[95] They further contended that this contamination occurred due to a seeming lack of enforcement of water quality as mandated by state law.[96] They argued that if the water quality laws were enforced as they were intended, the poultry companies would have been forced to offer alternative ways to dispose of poultry wastes while protecting the environment. The Lower Shore counties relied on income from the poultry industry to remain viable. On the other hand, the industry needed an outlet to dump its poultry waste at a cheap overhead cost or an incentive package to minimize the cost.[97]

Several of the poultry waste landfills were located in Berlin, Princess Anne, and Salisbury. The locations of most of these landfills were in the black communities and the decision to locate them there was not made by the poultry companies. Rather, local zonal ordinances created the enabling environments that designated these locations as either landfills for poultry wastes, or manufacturing sites for commercial poultry production.[98] In essence, the local political order aided the environmental transformation of African American communities on the Lower Maryland Shore in the sense that it permitted environmental pollution to occur. Local black leaders rallied against these zoning ordinances but fell short of effecting desired changes, because whites mainly controlled the votes in the local boards of county commissioners and supervisors.[99]

Environmentalists equally noted that the activities of commercial poultry producers contributed to air pollution. They stated that the health of people living in communities on the Shore was negatively impacted by the strong odor from both the chicken houses and the processing plants.[100] In particular, it was evident as this author drove through the black community in Salisbury and spoke to the residents that the unmistakable odor from the Perdue processing plant located right in the heart of the black community permeated through the air quality. This negatively impacted African Americans' quality of life because of the discomforting and consistently bad odor that emanated from the processing plant.

Indeed, a study conducted in 1990 by the Environmental Protection Agency (EPA) in California on odors from agricultural and industrial sites showed that those who lived near poultry farms, which were considered environmental odor sources (EOS), were likely to experience "aversive conditioning phenomena, stress-induced illness, and possible pheromonal reactions."[101] This study is relevant to the Lower Maryland Eastern Shore because of the concentration of poultry farms and the residents' exposure to the daily noxious smell. Various African American communities were particularly adversely affected since many of the processing plants and waste disposal sites were located either within their communities or close to their communities.[102] The issues of the impact of commercial poultry production on public health and the environment remained a major concern to environmentalists from the 1960s to the 1990s.

MARYLAND STATE GOVERNMENT'S REACTION TO THE PUBLIC HEALTH AND ENVIRONMENTAL PROBLEMS CAUSED BY COMMERCIAL POULTRY PRODUCTION

The concomitant effects of commercial poultry production on public health and the environment, led the State of Maryland to undertake a series of executive and legislative actions, to protect the public from the harmful effects of commercial poultry production. As public health and the environment became dominant issues of concern during the early phase of the commercialization of poultry, the Maryland legislature enacted legislations aimed at protecting the public from potential poultry-related disease and outbreaks of fish-killing *Pfiesteria*.

There were concerns about public exposure to adulterated and unsafe poultry in Maryland because of the state's emerging status as an important region for commercial poultry production. Against this backdrop, the Maryland legislature enacted legislation in 1937 that addressed public concerns with poor sanitation in slaughterhouses. In 1939, as the demand for poultry consumption rose, portions of the 1937 legislation were repealed. The new provisions included "obtaining a license from the State Board of Agriculture to engage in commercial poultry transactions, possessing a memorandum signed by the vendor and containing the vendor's address, date of sale, breed (in the case of live poultry), weight and number of poultry purchased or has such other information as will establish the ownership of same."[103]

The new provisions were partly in response to the popularity of commercial poultry in Maryland and the fact that the industry drew poultry brokers from the Northeastern United States to the state. There was a high probability that without the state exerting oversight, poultry consumers might be exposed to diseases resulting from unsanitary poultry. The requirement of brokers to obtain licenses from the State Board of Agriculture was therefore aimed at allaying public concerns over the probability of adulterated poultry.

As public concerns over the safety of poultry products lingered, the state legislature again repealed portions of the 1939 legislation in 1941, and reenacted it with amendments that provided for additional penalties for violators of poultry sanitation in Maryland. The new legislation empowered the Board of Agriculture to supervise all commercial poultry operations in Maryland.[104] Furthermore, in line with the federal Poultry Products Inspection Act of 1957, which required the continuous inspection of all poultry products in interstate commerce, the Maryland legislature enacted the Maryland Poultry Products Inspection Act in 1958. This Act further addressed the issue of poultry adulteration. It specifically defined certain actions such as preparing poultry in contaminated or unsanitary conditions, as well as the use of certain substances such as pesticides or color additives while preparing poultry, as violation of the inspection act.[105]

Towards the end of the 1950s, the federal government continued with its regulation and inspection of meat products but moved towards cooperative legislations with states. In 1958, the U.S. Congress enacted the Federal Meat Inspection Act, which was subsequently amended in 1967 as the Wholesome Meat Act. In 1968, it was amended yet again and became the Wholesome Poultry Products Act. This act as amended placed the responsibility of the inspection of processed meat within the Food Safety and Inspection Service (FSIS), an agency of the United States Department of Agriculture (USDA), and required states to adopt rigorous inspection standards.[106] Maryland then followed suit by enacting the Wholesome Meat Act in 1968 to monitor meat safety and the sanitary conditions in processing plants.[107]

By the 1970s, the federal and state governments established agencies and mechanisms for food inspection and safety in poultry slaughterhouses and processing plants. There were still safety issues that persisted in terms of sanitation and food contamination in the processing plants. Consequently states assumed responsibility for inspections of food producing facilities. In 1973, the Maryland Occupational Safety and Health (MOSH) was established and the agency was responsible for exercising oversight of commercial poultries' compliance with public health and environmental laws that related to commercial poultry production.[108] Also in 1973, the Maryland legislature re-enacted the Maryland Wholesome Meat Act with amendments, which included laws and regulations governing the licensing of poultry dealers, as well as sanitation in slaughtering establishments. The act provided instructions on the labeling of meat, including poultry, to ensure meat safety.[109] In some respect, the promulgation of this act was a way to further regulate and maintain high sanitation standards in poultry production in Maryland.

In the 1980s and in Maryland in particular, the major issues of concern were food contaminations and the state of the Chesapeake Bay. The public debate centered on finding the right balance between commercial poultry production, public health, and the environment. These public health and environmental issues notwithstanding, the Delmarva region remained an economic magnet in terms of the profitability of commercial poultry. Consequently, the governors of Delaware, Maryland, and Virginia established the Mid-Atlantic Poultry Health Council in 1987. The council mainly concerned itself with policy coordination in the event of a poultry disease outbreak. There was an economic motivation for the establishment of the council because all three states bore the brunt of any outbreak and suffered the most impact from a catastrophic poultry disease. The three states combined their resources to combat outbreaks when they occurred.[110]

In the 1990s, the Maryland legislature enacted a landmark legislation to address environmental pollution. In 1998, the Maryland legislature enacted the Water Quality Improvement Act (WQIA).[111] This Act imposed regulations on the amount of nutrients that were emitted into the Bay from poultry and other animal-related commercial production. Also, the Act required commercial poultry processors to use enzyme phytase to help with the digestibility of phosphorous found in chicken feed.[112] An important provision of the act was the establishment of an Animal Waste Technology Fund that required certain commercial feeds to contain certain ingredients for better nutrient management. The Act essentially excluded the use of poultry litter as fertilizer[113], which meant that commercial poultry producers had to find alternative uses for their poultry litter.[114] Poultry litter contain on the average 3.52 percent nitrogen, 2.97 percent phosphorous, and 2.34 percent potassium. Each ton of litter contains 70.44 pounds of nitrogen, 59.42 pounds of phosphorous, and 46.86 pounds of potassium.[115] These compounds contained toxins that when released were capable of killing fishes and other habitats. Perdue Farms alone allegedly generated 8–10 percent of the total quantity of poultry litter generated on the Delmarva Peninsula.[116]

Furthermore, there were a series of executive and legislative actions aimed at addressing the public health and environmental problems caused by the commercial poultry industry. In 1997, Governor Parris N. Glendenning appointed a Blue Ribbon Citizens *Pfiesteria* Action Commission made up of a broad segment of Marylanders. The Commission was charged with examining "the characteristics of the Pocomoke River and its watershed, and the characteristics of its watershed, to determine whether these waterways are uniquely vulnerable to toxic outbreaks of *Pfiesteria*. What is the likelihood of toxic outbreaks in these rivers in the future and elsewhere in Maryland? Is there a profile that we can apply to other watersheds in the State?"[117]

The Commission was then to make recommendations for minimizing the risk of further outbreak. The Commission's findings suggested that "toxins released by Pfiesteria in the Chesapeake Bay watershed and elsewhere can produce human consequences, primarily manifested as cognitive impairment, particularly impacting short-term memory abilities."[118] The intersection between *Pfiesteria* and commercial poultry occurred with runoffs from the processing plants that emptied into the tributaries that then connected to the Bay. Also at issue with the *Pfiesteria* outbreak was the way that the state managed its nitrogen by-products from commercial poultry production.

Prior to 1997, managing nitrogen was strictly voluntary. Companies were not mandated to follow specific rules for disposing nitrogen-based waste generated from commercial poultry processing. The Commission, after it had determined the role of commercial poultry in the outbreak, recommended that "the State [should] enroll all farmers in nutrient management plan by the year 2000."[119] Ironically, poultry waste has the highest content of phosphorous,[120] which perhaps explains why the Commission recommended it as part of the agricultural nutrient management program.

The General Assembly sought to add teeth to the Commission's recommendations by passing Senate Bill 178 and House Bill 599 in 1998. Both bills were intended to strike a balance between commercial poultry production, its importance to the local economy, and the environmental protection of the communities. The bills required farms using commercial poultry by-product like manure, which contained poultry fecal matter to develop and implement a nitrogen-based nutrient management plan by a given date, or pay fines of $250 imposed by the Maryland Department of Agriculture.[121]

During the 1998 Legislative Session, the State Senate defeated Senate Bill 413, which was intended to place a heavier burden on commercial poultry producers. The intent of the bill was to regulate the disposal of excess poultry waste by requiring the State Departments of Agriculture and the Environment to write new regulations governing how the poultry companies disposed of their waste and runoff from the processing plants, while protecting the environment.[122] Had the bill passed, it would have required the companies to pay for the cost of any environmental clean up by the State government. The failure to pass this bill in the Senate was indicative of the industry's lobbying influence in the Maryland General Assembly.

CONCLUSION

Issues of public health and the environment featured more prominently in the 1950s during the large-scale commercialization of poultry on the Lower Maryland Eastern Shore. Public health and the environment were of concern during the early phase of commercialization of poultry, but the wholesome approach to poultry production to meet a high demand from the 1950s and onward, increased the exposure to several diseases. The introduction of certain bacteria-killing medications and antibiotics raised yet more concern about the impact of these medications on consumers, public health, and the environment. However, as poultry production became more commercialized, the public health concern was inevitable because of the microorganisms that reside in animals. Bacteria such as *Campylobacter*, *Listeria*, *Salmonella*, *E. coli*, *Pseudomonas aeruginosa* are organisms that are commonly found in poultry. The presence of these bacteria in poultry meat raised concern for public health and the environment. Although researchers experimented with various drugs to attempt to minimize the persistence and outbreak of these microorganisms, the threat of the microorganisms to public health persisted.

The application of antibiotics was a major public health concern with commercial poultry production. While antibiotics were partially useful in treating some poultry-related diseases, their deliberate use to grow and fatten commercial chickens was a major public health concern. Silbergeld's protracted research and study has found that antibiotics used in commercial chickens allowed the proliferation of drug resistant microorganisms. These resistant strains posed a greater threat to public safety. They placed the public at a higher risk as more Americans consumed poultry meat and were therefore more likely to be exposed to poultry-related food poisoning.

In the environmental sphere, the major issues were water contamination, air pollution, and land degradation. Commercial poultry factories utilized millions of gallons of water in the processing plants. The water runoff from these factories was found to be contaminated with high-level amounts of algae and other toxins that killed fish upon contact and thereby exposed maritime fisheries to health risks. Another environmental concern was air pollution. Particularly, African Americans on the Lower Maryland Eastern Shore were adversely impacted by air pollution. The strong and protracted toxic odors that emanated from the processing plant in Salisbury and the poultry landfills in Somerset and Worcester Counties, which were located near African American communities, exposed them to dangerous air quality. Much of the large quantities of manure generated from the poultry houses in Somerset and Worcester Counties, contained nitrogen and phosphorous, which further polluted their environment.

The third major environmental concern was land degradation. Since the majority of Eastern Shore residents are farmers, the water runoff as well as excessive phosphorous present in the soil had a damaging effect. Crops farmers bore the brunt of this environmental degradation because of the nutrient management challenges. Although there was a mandated nutrient management program by the late 1990s, crops farmers of the previous years sought alternative ways to remove excess nutrients from their farms before cultivation of their crops. One alternative allowed farms to fallow for one or two seasons before farmers cultivated and harvested their crops. The outcome of this fallowing practice was lost income for many of the farmers.

Environmentally, commercial poultry production on the Lower Maryland Eastern Shore contributed to the increased amounts of nitrogen and phosphorous, which led to overgrowth of algae that released toxins into rivers and streams, which then emptied into the Bay, causing considerable negative impact to fishing and other maritime activities. Also, commercial poultry had a detrimental effect on the environment as a result of increased amounts of drugs in the soil, which stemmed from run-offs from poultry processing plants.

From a public health standpoint, three major factors in commercial poultry production included consumer safety issues such as exposure to food contamination in the food chain, and resistance to drugs used to treat poultry-related diseases that were also fed to chickens. That is, for the chickens to reach a certain size, they had to be fed specialized medications and hormones. Over time, these medications and hormones were overused in commercial poultry flock and increased the risk of human resistance to antibiotics for poultry diseases.

Overall, there was an intersection between commercial poultry production, public health, and the environment. This intersection was inevitable because of microorganisms that were present in commercially produced poultry. Essentially, commercial poultry production on the Lower Maryland Eastern Shore faced a major obstacle in maintaining a balance between a needed and important industry that was crucial to the economic life-wire of the region on one hand, and on the other, protecting public health and a sustainable environment.

NOTES

1. A very important work in this regard is Donald D. Stull and Michael J. Broadway, *Slaughterhouse Blues: The Meat and Poultry Industry in North America* (Belmont, California: Thomson/Wadsworth, 2004), which gave a historical analysis of the industrialization of meat production over the last century and half and its attendant social, environmental, and public health consequences. A few other works include, A.W. Hoadley, W.M. Kemp, A.C. Firmin, G.T. Smith, and P. Schelhorn, "*Salmonellae* in the Environment Around a Chicken Processing Plant," *Applied Microbiology*, Volume 27, No. 5 (May 1974), 848–857; B.J. Dutka and J.B. Bell, "Isolation of *Salmonella* from Moderately Polluted Waters," *Journal of Water Pollution Control Federation*, Volume 45, No. 2 (February 1973), 316–324; George K. Morris, Joy G

Wells, *Salmonella* Contamination in a Poultry-Processing Plant," *Applied Microbiology*, Volume 19, No. 5 (May 1970),795–799, and ANN Wilder, R.A. MacCready, "Isolation of *salmonella* from poultry, poultry products and poultry processing plants in Massachusetts," *New England Journal of Medicine*, Volume 274, No. 26 (June 30, 1966), 1453–1460.

2. For specific poultry-related diseases see, F.T.W. Jordan (ed.), *Poultry Diseases*, Third Edition, (London: Bailliere Tindall, 1990).

3. George W. Chaloupka, interview by author, Georgetown, Delaware. He retired as a poultry research scientist at the University of Delaware and conducted many experiments on respiratory diseases in commercial flocks. See also, David Sainbury, *Poultry Health and Management*, Second Edition (London: Granada Publishing Ltd., 1984).

4. Chaloupka interview.

5. Ibid. For a comprehensive listing of poultry diseases, see Mack O. North, *Commercial Chicken Production Manual*, Third Edition, (Westport, Connecticut: The AVI Publishing Inc., 1984), 612–668.

6. J. L. Skinner and M. L. Sunde, "Skinner and Sunde remember Poultry Digest's 50 years," *Poultry Digest* (October 1989), 444. See also George W. Chaloupka, "The Early Days of our Poultry," Georgetown, Delaware (n.d.). [This paper was donated to the author by Mr. Chaloupka from his private collections and is in the author's possession].

7. Skinner and Sunde, 444.

8. George W. Chaloupka, "The Early Days . . ."

9. Chaloupka, interview.

10. See Joshua Richard Hayes, "Multiple Antibiotic Resistances of *Enterococci* from the Poultry Production Environment and Characterization of the Macrolide-Lincosamide-Streptogramin Resistance Phenotypes of Enterococcus Faecium" (Ph.D. diss., University of Maryland, College Park, 2004).

11. Marcia Wood, "Toward a Safer Food Supply," *Agricultural Research* (December 1999). See also Agricultural Research Service Program (#108). Available at http://www.nps.ars.usda.gov/programs/appvs.htm.

12. See R. Reece Corey and Joseph M. Byrnes, "Oxytetracycline-Resistant Coliforms in Commercial Poultry Products," *Applied Microbiology*, Volume 11, No. 6 (November 1963), 481–484. See also the following related studies, M. E. Coates, C. D. Dickinson, G. F. Harrison, S. K. Kon, S. H. Cummins, and W. F. J. Cuthbertson, "Mode of Action of Antibiotics in Stimulating growth of Chicks," *Nature*, Volume 168 (1951), 332; H.R. Bird, R. J. Lillie, and J. R. Sizemore, "Environment and Stimulation of chick growth by antibiotics," *Poultry Science*, Volume 31 (1952), 907; H. Yacowitz, "Antibiotic levels in the digestive tract of the chick," *Poultry Science*, Volume 32 (1953), 966–8; G. R. Frye, H. H. Weise, and A. R. Winter, "Relative effectiveness of increasing shelf life of poultry meat by long and short periods of antibiotic feeding," *Food Technology* 12 [suppl], (1958), 52; R. F. Gordon, J. S. Garside, and J. F. Tucker, "Emergence of resistant strains of bacteria following the continuous feeding of antibiotic to poultry," *International Veterinary Congress, Madrid, Spain*,

Volume 2 (1959), 347–9, and M. S. Mameesh, B. Sass, and B. C. Johnson, "The assessment of the antibiotic growth response in the chick," *Poultry Science*, Volume 38 (1959), 512–515.

13. I am using "wrong" here in the sense that its application for anything other than solely treating diseases associated with chickens did not serve a public health interest. Instead, the antibiotics when used for making the chickens grow faster and bigger only served the interest of the commercial poultry companies because of the consumers' preference for meatier chickens on their dinner tables.

14. Quoted in Tom Pelton, "Poultry farms' use of antibiotics raises concerns about drug-resistant germs," *The Sun* (Baltimore), 31 August 2004.

15. "Science Overview: Public Health Implications of Industrial Animal Production," (n.d.) *Johns Hopkins Bloomberg School of Public Health, Baltimore, Maryland.* This School has on-going research examining both the impact of industrial animal production on the environment, public health, and specifically on the health of the workers in the industry on the lower Maryland Eastern Shore. See also, Leo Horrigan, Robert S. Lawrence, and Polly Walker, "How Sustainable Agriculture Can Address the Environmental and Human Health Harms of Industrial Agriculture," *Environmental Health Perspectives*, Volume 110, No. 5 (May 2002), 445–456. While many of the health, environmental and scientific experiments and studies cited throughout this work were conducted and published beyond the 1990s, several were actually longitudinal studies begun before the 1990s and the results published during or after the 1980s and 1990s. They showed the intersection between commercial poultry production, public health, and the environment from the 1930s through the 1990s. Also, many of the health and environmental issues raised during the early and large-scale phases of commercial poultry still exist. The major difference is that new medications have been found that cured many of the diseases that were more prevalent during the early phase of commercial poultry production.

16. "Science Overview: Public Health Implications of Industrial Animal Production . . ."

17. Ibid.

18. For a more focused discussion on antibiotic and its early beginnings, see Brian James Gangle, "Sources and Occurrence of Antibiotic Resistance in the Environment," (M.S. thesis, University of Maryland, College Park, 2005), especially chapter 1.

19. Tom Pelton, "Poultry farms' use of antibiotics raises concerns about drug-resistant germs," *The Sun* (Baltimore), 31 August 2004. See also other related studies, "Bacteria May be Linked to Poultry Worker Illness," http://www.keepantibioticworking.com; "Risk Assessment on the Human Health Impact of Fluoroquinolone Resistant Campylobacter Associated with the Consumption of Chicken," Food and Drug Administration, Center for Veterinary Medicine, 5 January 2001 at http://www.fda.gov/cvm; "Prescription for Trouble: Using Antibiotics to fatten Livestock," http://www.ucsusa.org/food, among others. The FDA article was based on experiments from stool samples taken from those who contracted campylobacter after consuming chicken in 1998. The Union of Concerned Scientists' (UCS) article generally argued

against using growth hormone for livestock particularly and believed that antibiotics can only be used for therapeutic purposes.

20. U.S. Public Law 85-172, Title 21—Food And Drugs, Chapter 10—Poultry And Poultry Products Inspection Act of 1957.

21. See Donald D. Stull and Michael J. Broadway, *Slaughterhouse Blues . . .*

22. See Homer W. Walker and John C. Ayres, "Incidence and Kinds of Microorganisms Associated with Commercially Dressed Poultry," *Applied Microbiology*, Volume 4, No. 6 (November 1956), 345–349, D. J. Kraft, Carolyn Olechowski-Gerhardt, J. Berkowitz, and M. S. Finstein, "Salmonella in Wastes Produced at Commercial Poultry Farms," *Applied Microbiology*, Volume 18, No. 5 (November 1969), 703–707, and Judith L. Aulik and Arthur J. Maurer, "Lactic Acid Bacteria in Poultry Products: Friend or Foe, " *Poultry and Avian Biology Reviews*, Volume 6, No. 3 (1993), 145–184.

23. See the following studies, R.R. Kazwala, J.D. Collins, J. Hannan, R.A. Crinion, and H. O'Mahony, "Factors responsible for the introduction and spread of *Campylobacter jejuni* infection in commercial poultry production," *Veterinary Record*, Volume 126, No. 13 (March 1990), 305–6; Doris Stanley, "Steaming out the *Salmonella* risk-destroying *Salmonella* and other poultry microorganisms," *Agricultural Research* (October 1997); Thomas P. Oscar, "The Development of a Risk Assessment Model for Use in the Poultry Industry," *Journal of Food Safety* 18 (1998), 371–381; Donald Corrier, "Preemptive Strike Against *Salmonella*," *Agricultural Research* (June 1998); Marcia Wood, "Fast Tests for *Campylobacter*," *Ibid.* (March 1999); S. M. Shane, "*Campylobacter* infection in commercial poultry," *Revue Scientifique et Technique*, Volume 19, No. 2 (August 2000), 367–395; K. S. Yoon and T. P. Oscar, "Survival of Salmonella typhimurium of Sterile Ground Chicken Breast Patties After Washing with Salt and Phosphates and During Refrigerated and Frozen Storage," *Journal of Food Science*, Volume 67, No. 2 (2002), 772–775; S. Datta, H. Niwa, and K. Itoh, "Prevalence of 11 pathogenic genes of Campylobacter jejuni by PCR in Strains isolated from humans, poultry meat and broiler and bovine faeces," *Journal of Medical Microbiology*, Volume 52, No. 4 (April 2003), 345–348; D.G. Newell, and C. Fearnley, "Sources of Campylobacter Colonization in Broiler Chickens," *Applied and Environmental Microbiology*, Volume 69, No. 8 (August 2003), 4343–4351; K.S. Yoon, C. N. Burnette, and T. P. Oscar, "Development of Predictive Models for the Survival of Campylobacter *jejuni* (ATCC 43051) on Cooked Chicken breast Patties and in Broth as a Function of Temperature," *Journal of Food Protection*, Volume 67, No. 1 (2004), 64–70; T. P. Oscar, "Development and Validation of Primary, Secondary, and Tertiary Models for Growth of Salmonella Typhimurium in Sterile Chicken," *Journal of Food Protection*, Volume 68, No. 12 (2005), 2606–2613; T. P. Oscar, "Validation of Lag Time and Growth Rate Models for Salmonella Typhimurium: Acceptable Prediction Zone Method," *Journal of Food Science*, Volume 70, No. 2 (2005), 129–137, and T. P. Oscar, "Simulation Model for Enumeration of Salmonella of PCR Detection Time Score and Sample Size: Implications for Risk Management," *Journal of Food Protection*, Volume 67, No. 6. (2004), 1201–1208.

24. Jones, Lu Ann and Nancy Grey Osterud, "Breaking New Ground: Oral History and Agricultural History," *Journal of American History* Volume 76 (September 1989), 556. Several studies have also linked respiratory problems to the dust and ammonia from the chicken houses. Although some of the studies were conducted in the last ten to fifteen years, the effects of the ammonia and dust on the workers during the 1930s and 1940s were the same as in subsequent years. The health effects on poultry workers in other parts of the country also affected those workers on the Lower Maryland Eastern Shore. See the following studies, Kenneth Rosenman, "Occupational Asthma and Farming," *Physician's Newsletter* Volume 2, No.2 (Winter 1993); Dennis J. Murphy and Cathleen M. LaCross, "Farm Respiratory Protection," *Fact Sheet Safety 36*, Pennsylvania Cooperative Extension Service, 1993; P. Morris, et al., "Respiratory Symptoms and Pulmonary Function in Chicken Catchers in Poultry Confinement Units," *American Journal of Industrial Medicine*, Volume 19 (1991), 195–204; Ronald C. Jester and George W. Malone, "Respiratory Health on the Poultry Farm," Available at http://www.cdc.gov/nasd/docs, and http://www.ufcw.org/workplace.

25. Several former chicken catchers that were interviewed indicated their long-term struggle with respiratory infections resulting from years of working inside the confined chicken houses. See also Peter S. Goodman, "Eating Chicken Dust," *The Washington Post*, 28 November 1999.

26. Jacek Dutkiewicz, "Exposure to Dust-Borne Bacteria in Agriculture. I. Environmental Studies," *Archives of Environmental Health*, Volume 33, Issue 5 (September/October 1978), 250–259.

27. Muneo Wakabayashi, Betsy G. Bang, and Frederick B. Bang, "Mucociliary Transport in Chickens with Newcastle Disease Virus and Exposed to Sulfur Dioxide," *Archives of Environmental Health,* Volume 32, Issue 3 (May/June 1977), 101–108.

28. Cited in Brice Stump, "Poultry House Dust Causes Health Problems," *Daily Times*, 23 February 1987.

29. See Peter S. Goodman, "Eating Chicken Dust . . ."

30. Brice Stump, "Litter Research Has Poultry Specialist Excited," *Daily Times*, 26 September 1988.

31. Marcia Wood, "Cleaner chicken ahead—poultry processing," *Agricultural Research* (September 1994).

32. Leslie E. Card and Malden C. Nesheim, *Poultry Production* (Philadelphia: Lea and Febiger, 1966), 245–276.

33. See the following related studies by, Colin G. Scanes, "Introduction: Chickens—A Model for Growth?" *Critical Reviews in Poultry Biology*, Volume 3, No. 4 (1991), 225–227; R. E. Ricklefs, "Modification of growth and development of muscles of poultry," *Poultry Science*, Volume 64 (1985); H.D. Chapman, Z.B. Johnson, and J.L. McFarland, *Poultry Science*, Volume 82 (2003), 50–3, and S Harvey, R.A. Fraser, and R.W. Lea, "Growth hormone secretion in Poultry," *Critical Reviews in Poultry Biology*, Volume 3, No. 4 (1991), 239–282.

34. See the following works, Sandy Miller Hays, "Detecting poultry drug residues," *Agricultural Research* (January 1994), and *Ibid*, "Two strategies for protecting poultry from Coccidia," *Ibid* (October 1996).

35. Quoted in Keith Schneider, "F.D.A. Faulted in Threat from Animal Drugs," *The New York Times*, 13 January 1986.

36. Marian Burros, "Fact and Fancy on the New Low-Fat Chickens," *The New York Times*, 26 February 1986.

37. John L. Skinner, *American Poultry History, 1974–1993*, Vol. II (Mount Morris, Illinois: Watt Publishing Company, 1996), 112.

38. Karen Davis, *Prisoned Chickens, Prisoned Eggs: An Inside Look at the Modern Poultry Plant* (Summertown, Tennessee: Book Publishing Company, 1996), 102–104.

39. Ibid.

40. See the following, Don Kendall, "Nearly 4 in 19 Chickens May be Salmonella-Tainted," *The Washington Post*, 18 February 1987; Patricia Picone Mitchell, "Avoiding Salmonella Contamination," *The Washington Post*, 18 March 1987; Ward Sinclair, "Poultry Contamination Undetected, Report Says," *The Washington Post*, 13 May 1987; "Contaminated Inspection Process," *The Washington Post* [Op/Ed], 25 May 1987; Ward Sinclair, "Poultry Inspection Flaws Admitted," *The Washington Post*, 3 June 1987; "Meat and Poultry: Inspection in Modern Times," *The Washington Post* [Op/Ed], 20 June 1987; Patricia Picone Mitchell, "The New Consumer Crusade: Momentum Builds in the Fight Against Contaminated," *The Washington Post*, 16 September 1987, and Marian Burros, "Eating Well," *The New York Times*, 7 December 1988.

41. "FDA Approves Antibiotic for Chickens: Drug Aimed at Preventing the Spread of Deadly E. Coli Bacteria," *The Washington Post*, 19 August 1995.

42. See the following, Marian Burros, "FDA Approves the Irradiation of Poultry," *The New York Times*, 2 May 1990; Food and Drug Administration, "Irradiation in the production, processing, and handling of food," [Final rule], *Federal Register*, 55, (May 1990), 18538–18544; Food Safety and Inspection Service, "Irradiation and poultry products," [proposed rule], *Federal Register*, 57, (September 1992), 43588–43600; Food and Drug Administration, "Irradiation in the production, processing, and handling of food," [Final rule], *Federal Register*, 60, (March 1995), 12669–12670, and Food and Drug Administration, "Irradiation in the production, processing, and handing of food," [Final rule], *Federal Register*, 62, (December 1997), 64107–64121.

43. Marian Burros, "FDA Approves the Irradiation of Poultry . . ."

44. See Alan Ismond and P. Eng, "Irradiation's not the Solution," *Letters, Meat & Poultry*, (December 1995), 4.

45. See the following, Carole Sugarman, "A Promising Solution to Salmonella," *The Washington Post*, 12 August 1992.

46. Marian Burrows, "U.S. Approves Chicken Treatment to Cut Salmonella," *The New York Times*, 14 October 1992.

47. Ibid, "A New Spray for Chickens Helps Control Salmonella," *The New York Times*, 20 March 1998.

48. Martin Tolchin, "Clinton Orders Hiring of 160 Meat Inspectors," *The New York Times*, 12 February 1993.

49. Ibid.

50. Marian Burrows, "Eating Well: U.S. Eases Up on Irradiation, Antibiotics," *The New York Times*, 26 August 1998.

51. "Intensive Poultry Production: Fouling the Environment," Available at http://www.upc-online.org. See also the following works, Kim Clark, "Chicken manure fouls the Bay," *The Sun* (Baltimore), 21 March 1993; Senator Tom Harkin, "Animal Waste Pollution in America: An Emerging National Problem." (A Report compiled by the Minority Staff of the Committee on Agriculture, Nutrition, and Forestry, December 1997); J. Gerstenzang, "Poultry Production Threatens Potomac River's Health," *San Francisco Chronicle*, 21 April 1997; "The Poultry Industry and Water Pollution in the South," A Report, *Institute for Southern Studies*, Durham, North Carolina, December 1990, and J. Warrick and T. Shields, "Maryland Counties Awash in Pollution-Causing Nutrients," the *Washington Post*, 3 October 1997.

52. There is no definitive beginning point of environmental concerns regarding the Chesapeake Bay and its tributaries. The health of the Bay was also a matter of interest to the State in general and to those who had to rely on the healthiness of the Bay to make their living. However, the era of large-scale commercialization of poultry production in the 1950s, marked the beginning of the increased amounts of ammonia, for example, and other poultry-related contaminants flowing into the Bay and its environs. Sustained environmental studies began appearing around the 1970s after the Chesapeake Bay Foundation was founded in 1967. The Foundation eventually became the preeminent environmental group solely dedicated to the preservation of the Bay and its environs. The Foundation also formally began tracking activities on the Bay including effects of commercial animal production and its impact on the health and life of the Bay. It commissioned and supported scholarly studies about the Bay. Published articles about the Bay became more available from the 1980s and thereafter.

53. See Michael Gochfeld and Bernard D. Goldstein, "Lessons in Environmental Health in the Twentieth Century," *Annual Review of Public Health*, Volume 20, Issue 1 (1999), 35–53.

54. Henry S. Parker, "Agriculture and Marine Environments," *Agricultural Research*, (January 1999), and Don Collins, "Protecting the Chesapeake," *Agricultural Research*, (January 1999).

55. The NIMBY Syndrome originated in the 1980s from the environmental movement and preservationists who argued against environmental degradation and other developmental actions of the government that were deemed unfair or detrimental to the wishes of localities. When used in an environmental context, it's a "fighting and rallying" phrase by those who wish to force the government to manage public waste in a more environmentally-friendly way. See the following works by Debra Stein, "NIMBYism and Conflict of Interest," *Public Management Magazine* (August 2006); "The NIMBY Report," *National Low income Housing Coalition* (February 2001); "The Ethics of NIMBYism," *Journal of Housing and Community Development* (November/December 1996), and *Winning Community Support for Land Use Projects* (Washington, D.C.: Urban Land Institute, 1992).

56. See Kristen Chossek, Polly Walker, Thomas A. Burke, and Beth Resnick, "The Chesapeake Bay Health Indicators Project: Linking Ecological and Human

Health," (The Center for a Livable Future, Johns Hopkins Bloomberg School of Public Health, Baltimore, Maryland) [n.d.].

57. William C. Baker and John D. Groopman, "Introduction: Health of the Bay—Health of People Colloquium," *Environmental Research Section* A 82, (2000), 95–96. This was a meeting held in 1998 to explore the links between human health and the ecological system. Although this meeting was held in the 1990s, it did not mean that there had not been concerns about commercial poultry production and the ecological systems prior to the 1990s. The concern was always present. The difference was in the new technologies that have been developed to better study the inextricable links between commercial poultry production and the environment. Also, as poultry production became a much larger enterprise during the vertical integration period, its broader implications on public and environmental health became more prominent.

58. See Donald F. Boesch, "Measuring the Health of the Chesapeake Bay: Toward Integration and Prediction," *Environmental Research Section* A 82 (2000), 134–142.

59. See the following related studies on the Chesapeake Bay, R.A. Eskin, K. H. Rowland, and D. Alegre, "Contaminants in the Chesapeake Bay Sediments 1984–1991," Chesapeake Bay Program CBP/RRS 145/96 (1996), Annapolis, Maryland; J. Greer and D. Terlizzi, "Chemical Contamination in the Chesapeake Bay: A Synthesis of Research to Date and Future Research Directions," (College Park: Maryland Sea Grant College, 1997); and "Targeting Toxins: A Characterization Report," Chesapeake Bay Program, 1999, Annapolis, Maryland, and Peter L. deFur and Lisa Foersom, "Toxic Chemicals: Can What We Don't Know Harm Us?" *Environmental Research Section* A 82 (2000), 113–133.

60. Tara Weaver, "Managing Poultry Manure Nutrients," *Agricultural Research*, (June 1998).

61. Robbin Marks and Rebecca Knuffke, *America's Animal Factories: How States Fail to Prevent Pollution from Livestock Waste,* (New York: Natural Resources Defense Council and the Clean Water Network, 1998), 49–51. See also, http://www.chesapeakebay.net/bayprogram.

62. David Morris and Jessica Nelson, "Looking Before We Leap: A Perspective on Public Subsidies for Burning Poultry Manure," *Institute for Local Self-Reliance*, October 1999. Also available at http://www.carbohydrateeconomy.org/library.

63. See the following, Sandy Miller Hays, "A Cleanup for Poultry Litter," *Agricultural Research*, (May 1994), and Doug Parker, "Alternative Uses for Poultry Litter," *Economic Viewpoints*, Volume 3, No. 1 (Summer 1998) among others. Available at http://www.arec.umd.edu/agnrpolicycenter. See also, A. Sharpley, *Agricultural Phosphorous in the Chesapeake Bay Watershed: Current Status and Future Trends* (Annapolis, Maryland: Scientific and Technical Advisory Committee, Chesapeake Bay Program, 1998).

64. I am using formalized ways here in the sense that there were no definitive laws in Maryland that pointedly mandated commercial poultry producers during the early phase of commercial poultry production to adopt specific waste management practices. However, as both the federal and state governments took on more active roles in regulating the industry, sanitation standards, which required slaughter houses and processing plants to be free of poultry debris, meant that commercial poultry

producers had to follow certain guidelines to dispose of their poultry wastes. Some of the local residents that were interviewed also indicated to this researcher that it was typical for poultry producers to truck their wastes to the landfill where they were emptied. It was not until the 1990s after the outbreak of *Pfiesteria Piscicida,* which affected fish life on several rivers of the Lower Eastern Shore that scientists made a link between commercial poultry waste and the outbreak. After this incident, the Maryland Legislature passed laws that made it mandatory for commercial poultry producers to follow waste management practices.

65. T.W. Simpson, "Agronomic Use of Poultry Industry Waste," *Poultry Science* 70 (1991), 1126–1131.

66. Available at http://www.perdue.com/company/committments/stewardship .html.

67. See Larry Goff, "Environmental Issues Are Concerns of Integrators, Growers," *Poultry and Egg Marketing*, (November/December 1997), 22.

68. See L. E. Cronin, *Pollution in Chesapeake Bay: A Case History and Assessment. Impact of Man on the Coastal Environment* (Washington, D.C.: U.S. Environmental Protection Agency, 1982), 17–46, and J. Capper, G. Power, and F.R. Shivers, Jr., *Chesapeake Waters: Pollution, Public Health, and Public Opinion, 1607–1972* (Centerville, Maryland: Tidewater Publishers, 1983) among other works.

69. United States Code, Public Law Title 42 [The Public Health and Welfare], Chapter 103, Comprehensive Environmental Response, Compensation, and Liability Act of 1980.

70. United States Environmental Protection Agency, "Responses to Comments on the Notice of Proposed Rulemaking on Superfund Notification Requirements and the Adjustment to Reportable Quantities," (Washington, D.C.: Environmental Protection Agency, 1985). See also, "Notice of Availability of a Petition for Exemption from EPCRA and CERCLA Reporting Requirements for Ammonia from Poultry Operations," *Federal Register*, Volume 70, Number 247 (December 2005); Available at http://wais.access.gpo.gov. This notice was in response to a petition filed by the poultry industry seeking exemption from the reporting requirements of two key environmental legislations that affected the commercial poultry industry, which were enacted in the 1980s.

71. See, *State of the Bay* yearly reports, 1980–1990. For example, during the 1980s, the Chesapeake Bay Foundation maintained a yearly report on the health of the Bay. In each of the reports, commercial poultry production was consistently cited as partially responsible for environmental degradation because of runoffs from the poultry factories into streams and creeks that emptied into the Bay.

72. There had been governmental intervention in the past with regard to environmental laws. However, the new laws that were enacted by Congress as a consequence of these reports emboldened communities across the country in protesting against the adverse environmental impact of commercial poultry production.

73. United States Code, Public Law Title 42 [The Public Health and Welfare], Chapter 116, Emergency Planning and Community Right to Know Act of 1986.

74. United States Code, Public Law Title 42 [The Public Health and Welfare], Chapter 133, Pollution Prevention Act of 1990. See http://www.epa.gov/ppa.htm.

75. See http://toxnet.nlm.nih.gov/cgi-bin/sis/htmlgen?TRI [Toxicology Data Network], TRI87-95 reports. Although the company's report indicated that no chemicals were released or transferred in terms of air pollution, it is plausible that these toxic chemicals found their way into the environment through poultry waste products and other means.

76. Peter S. Goodman, "Perdue to Help Farmers Dispose Chicken Waste: Company Planning to Turn Tons of Manure into Fertilizer and Sell It Elsewhere," *The Washington Post*, 25 February 1999.

77. Ibid, "Poultry Firms Agree on Plan to Limit Pollution," *The Washington Post*, 10 December 1998.

78. See the following, Eugene L. Meyer, "Maryland Chicken Farmers Reject Bird's Link to Fish Kills," *The Washington Post,* 14 September 1997; Todd Shields, "Poultry-Pfiesteria Link Challenged: Few Chicken Farms Near One Infested River, Maryland Officials Say," *The Washington Post*, 27 September 1997; Peter S. Goodman, "Manure Mushrooms into Mountain of a Problem: As Maryland Panel Studies the Hazards of Using Chicken Waste as Fertilizer, Alternative Uses Sought," *The Washington Post*, 3 October 1997; Peter S. Goodman, "Poultry Industry Urges Maryland to Delay Restrictions on Animal Waste," *The Washington Post*, 10 October 1997; Peter S. Goodman, "Poultry Limits Rejected: Maryland Microbe Panel Votes After Hot Debate," 28 October 1997, and Peter S. Goodman, "Poultry Group Offers $1 Million for Study," *The Washington Post*, 30 October 1997.

79. The Humane Society of the United States, "Pfiesteria and the Factory Farm," (a video documentary, 1998). See also, Thomas V. Grasso, "Pfiesteria fight may not be over: Big Poultry Firms must be held accountable for Chesapeake's health," *The Baltimore Sun* [Perspective], 17 May 1998; Michael Janofsky, "Fears of Deadly Organism Cast Shadow on Chesapeake," *The New York Times*, 26 April 1998.

80. "Corporate Hogs at the public Trough: Perdue Farms, Maryland," Available at http://www.sierraclub.org/factoryfarms/report99/perdue.asp.

81. Todd Shields, "Investigators to Focus on Runoff from Farms," *The Washington Post*, 8 September 1997.

82. Tyson Foods Inc., Annual Report, 1997, p.6.

83. David Segal, "Hudson Foods Has 'Long History' of Safety, Health Violations, OSHA Says," *The Washington Post*, 23 August 1997.

84. Amy Argetsinger, "Maryland Files Suit Against Perdue Farms," *The Washington Post*, 16 March 1997.

85. Tyson Foods Inc., Annual Report, 1997, p.6.

86. "Intensive Poultry Production: Fouling the Environment," Available at http://www.upc-online.org/fouling.html. See also, M. James, "Poultry plant to pay $6 million for polluting," *The Sun* (Baltimore), 8 May 1998.

87. Patrick Harmon, interview by author, Pocomoke City, Maryland. Mr. Harmon worked at this plant and other processing plants until he was injured on the job. He worked in the poultry industry for eight years in various capacities including stints as a chicken catcher. He also worked as an outreach/organizer at the Delmarva Poultry Justice Alliance to educate his former co-workers about the dangers and long-term effects of inhaling poultry dust, a consistent reality for chicken catchers. He focused on

mobilizing catchers to organize for better wages/compensations, working conditions, and respect, among other issues.

88. Available at http://www.cbf.org.

89. Ibid.

90. Michael H. Shuman, *Bay Friendly Chicken: Reinventing the Delmarva Poultry Industry* (Maryland: The Chesapeake Bay Foundation and the Delmarva Poultry Justice Alliance, 2000), 12–20.

91. Vincent Wilson, Jr., *The Book of the States*. (Brookeville, Maryland: American History Research Associates, 1992): 46.

92. See Michael H. Shuman, *Bay Friendly Chicken . . .*

93. See Gary L. Miller and John H. Axley, "Poultry Processing Plant Wastewater Disposal on SOD," Technical Report No. 36, [Completed Report A-013-Md 14-31-0001-5020], *Water Resources Research Center*, University of Maryland, College Park (July 1969–1975), 74. See also the following relevant studies, H. Bouwer, "Returning Wastes to the Land: A New Role For Agriculture," *Journal of Soil Water Conservation*, Volume 23 (1968), 164–168; W.E. Bullard, Jr., "Natural Filters for Agricultural Wastes," *Soil Conservation*, Volume 34 (1968), 75–77, and V. Larsen, "Agricultural Wastewater Rectification from a Poultry Processing Plant," (M.S. thesis, University of Maryland, College Park, 1970).

94. Peter S. Goodman, "A Look Inside the Modern Poultry Plant," *The Washington Post*, 2 August 1999.

95. J. Glen Morris, Jr., "Learning and Memory Difficulties after Environmental Exposure to Waterways Containing Toxin-Producing *Pfiesteria* or *Pfiesteria*-like Dinoflagellates," *The Lancet*, Volume 352, Issue 9127 (15 August 1998), 523–539.

96. Paul L. Sorisio, "Poultry, Waste, and Pollution: The lack of Enforcement of Maryland's Water Quality Improvement Act," *Maryland Law Review* 62, No. 4 (2003): 1054–1075.

97. Anna Smith, interview by author. Mrs. Smith has lived in the Berlin community for more than half a century.

98. See Charters of Somerset, Wicomico and Worcester Counties. Each county had its own zoning codes and regulations. Making these zoning decisions required a balance between county-wide interests and those of individuals and communities. Due to its long history of racial tension, it is plausible that race might have been a factor in making zoning decisions that disproportionately placed landfills either within or near black communities.

99. Diana Purnell, interview by author, Berlin, Maryland.

100. Hattie Beckwith, Eremin Stoudmire, and Alison Morton, interviews by author.

101. Dennis Shusterman, "Critical Review: The Health Significance of Environmental Odor Pollution," *Archives of Environmental Health*, Volume 47, No. 1 (January/February 1992), 76–87.

102. In Somerset and Worcester Counties, the poor air quality was evident because of persistent odors from chicken farm houses. This poor air quality that resulted from commercial poultry production was an added stress to the daily routine of the local residents. The odor was strong year round as many of the growers had the responsibil-

ity to dispose their poultry waste. For the growers who did not have the money to have them sent off site and were unable to convert the entire waste to fertilizer, they bore the intense odor and seemed to have gotten used to the odor. Finding a way to manage the odor was an added burden that created tension between the affected communities and commercial poultry producers.

103. See Annotated Code of Maryland, Article 48, Section 1, Chapters 147–152. Available at http://www.mdarchives.state.md.us/megafile/msa/speccol

104. Annotated Code of Maryland, Article 48, Section 1, Chapter 168, 1939 [1941].

105. See Annotated Code of Maryland, Agriculture Article, Title 4, subtitle 2. Also available at http://mlis.state.md.us/cgi-win/web_statutes.exe?gag&4-201.

106. Available at http://www.fsis.usda.gov/About_FSIS/100_Years_FMIA/index.asp.

107. See Annotated Code of Maryland, Agriculture, Title 4. Regulation of Livestock, Poultry Products and Eggs. Subtitle 1. Maryland Wholesome Meat Act of 1968. Available at http://www.mdarchives.com/msa/specol.

108. Available at http://www.dllr.state.md.us/mosh.

109. Annotated Code of Maryland, Agriculture, Subsection 4-101-131, Title 4, Regulation of Livestock, Poultry Products, and Eggs. Subtitle 1, Maryland Wholesome Meat Act, 1973. Available at http://www.dsd.state.md.us/comar/Annot_Code_Idx/AGIndex.htm.

110. See http://www.mdarchives.state.md.us/megafile/msa/speccol.

111. General Assembly of Maryland, Senate Bill 178/House Bill 599, 1997 Legislative Session. See also, Maryland Cooperative Extension, *A Citizen's Guide to the Water Improvement Act of 1998* (Annapolis, Maryland: University of Maryland, 1998).

112. See *Economic Situation and Prospects for Maryland Agriculture*, Policy Analysis Report No. 02-01 (College Park: University of Maryland, Center for Agricultural and Natural Resources Policy), 50.

113. See "An Assessment of the Benefits of Building FibroShore: Green Energy from Poultry Litter and Forestry Residues," A Report by the *Atlantic Resource Management, Inc.*, January 2002. Available at http://www.fibrowattusa.com/US-FibroShore/Assessment.

114. Erick Lichtenberg, Doug Parker, and Lori Lynch, "Economic Value of Poultry Litter Supplies In Alternative Uses," Policy Analysis Report No. 02-02, *Center for Agricultural and Natural Resource Policy*, University of Maryland, College Park, 2002.

115. Ibid, 7.

116. Ibid, 29. See also, Tom Ventsias, "Waste Not, Want Not: Putting Delmarva's Poultry Litter to Good Use," *Maryland Research*, Volume 11, Number 2 (Spring 2002). Available at http://www.marylandresearch.umd.edu/issues/spring2002/feature.html.

117. Report of the Governor's Blue Ribbon, Citizens *Pfiesteria Piscicida* Action Commission, Annapolis, Maryland, (November 3, 1997), 37 passim, [hereafter

Commission Report]. See also, "*Pfiesteria* and the Factory Farm," Human Society of the United States [video documentary, n.d. and in the author's possession].

118. Commission Report, 25.

119. *1998 90 Day Report*, [Agriculture: Nutrient Management], K–9; Education, Health, and Environmental Affairs Committee, Maryland State Senate, Annapolis, Maryland.

120. Vaclav Smil, "Phosphorous in the Environment: Natural Flows and Human Interferences," *Annual Reviews of Energy and the Environment*, 25 (2000), 63.

121. See the *1998 90 Day Report.*

122. Ibid, K–11.

Chapter Six

Commercial Poultry Production, Working Conditions, and Labor Activism

As commercial poultry production became a large-scale enterprise, it contributed to and enhanced Maryland's economy in general and the economy of the Lower Eastern Shore in particular. Maryland was continuously ranked among the top five to ten regions for commercial poultry production in the United States for most of the 1950s through the 1990s.[1] Vertical integration, consolidation, and research and development, as well as innovations were important factors that contributed to Maryland being a major player in commercial poultry production. However, the experiences of the poultry working class on the Lower Eastern Shore mirrored the experiences of workers in similar labor intensive occupations.[2]

Maryland's position as a major commercial poultry producer would not have been possible without the labor and participation of individuals who worked under challenging conditions. African Americans constituted a significant segment of these working individuals. The working conditions ranged from long hours without breaks, to no workman's compensation coverage for the workers. Workers faced major risks to their health and safety such as physical injury and even death as they worked on the assembly lines. There were instances where workers were permanently disabled as a result of injuries sustained while operating plant machines.[3]

The working conditions in the poultry houses and the processing plants became the focus of the labor activism that emerged particularly in the 1980s and 1990s. Labor activists and advocacy groups protested and demanded better working conditions. The protestations and demands led to the emergence of multi-ethnic labor activism, which involved whites, African Americans, Hispanics, Asians, and Caribbean workers. Prior to the arrival of Hispanics on the Lower Eastern Shore in the 1970s, African Americans constituted the bulk of the poultry labor force. However, poultry labor activism did not become prominent until the late 1980s through the 1990s.

The push for change in the working conditions in the commercial poultry industry was widespread. California, Arkansas, Georgia, and North Carolina all faced similar criticisms about the working conditions from labor unions and human and workers' rights organizations as did commercial poultry production on the Lower Maryland Eastern Shore. Working conditions in the poultry and meat industry in the South essentially mirrored the working conditions in the poultry processing plants on the Lower Maryland Eastern Shore where workers faced multiple risks, and exposure to injuries. The poultry companies' tendency to provide minimum safety measures for workers both in the poultry houses and processing plants was documented in a report by the Human Rights Watch.[4]

As workplace accidents in the commercial poultry industry became common place due to a lack of adequate safety measures, the Occupational Safety and Health Administration (OSHA) of the U.S. Department of Labor instituted regulations that required poultry companies to provide better safety measures as well as safer working conditions for their workers.[5] One measure included making protective wear accessible to workers and requiring line managers and supervisors to allow the workers time to put on this gear before they embarked on their daily processing line activity. The establishment of OSHA set the tone for workplace safety around the country as states established agencies to monitor compliance.

Maryland followed suit by enacting the Maryland Occupational Safety and Health Act of 1973. This act created the Maryland Occupational Safety and Health (MOSH), an agency within the Department of Labor, Licensing, and Regulations (DLLR). The act provided for job safety and health protection for workers through the promotion of safe and healthful working conditions throughout the State. Its mission was to "promote and assure workplace safety and health and reduce workplace fatalities, injuries, and illnesses."[6] Prior to the 1970s, it was common for fatalities, injuries, and illnesses in the poultry processing plants to be either unreported or under-reported. However, the establishment of OSHA and MOSH was indicative of the widespread violations of safe working conditions and labor rules, regulations, and laws. Before their creation, there was no means by which to require safety standards to be strictly enforced. These agencies were created due to the lack of strict enforcement of workplace safety laws.[7] The combination of lax enforcement of labor laws and unsafe working environments gave impetus to labor activism.

COMMERCIAL POULTRY PRODUCTION AND WORKING CONDITIONS

The working conditions in the 1930s and 1940s were harsh. The U.S. Bureau of Labor Statistics released a report in 1939 in which it estimated the number of employees in the manufacturing sector in the Northern part of the U.S. (of which Maryland was a part), who were wage earners and subject to the Fair Labor Standards Act to be 5,285,000 workers and only 632,500 as salaried employees. Agriculture and poultry processing were classified as part of the manufacturing industry.[8] Workers that fell into the manufacturing category were ineligible to receive comprehensive benefit package such as bonuses, health insurance, and workman's compensation as the salaried employees received.

On the Lower Maryland Shore, the lack of opportunities resulting from lack of education among majority of the residents, particularly African Americans, created a large pool of wage laborers from which the poultry industry drew. The low educational achievement of the workers was beneficial to the poultry companies because poultry processing and chicken catching did not require expensive training or company investment. Implicitly, by spending fewer resources, the companies reaped larger profits due to low overhead because of the depressed wages paid to workers.[9]

As poultry production became more commercialized, an underclass of mainly less educated residents emerged and helped sustain the new industry. Swift and Company, one of the earliest poultry processing plants that began operating in Salisbury in 1940,[10] benefited from the large pool of less educated workers who were in desperate need of work. In 1945, Southern States Eastern Shore Market Co-operative built a processing plant in Salisbury and hired workers with a starting wage of 35–40 cents per hour. The commercial poultry labor workers were wage earners, and their status as the underclass was tied to their limited access to education. Their jobs included killing, eviscerating, and icing down chickens before they were sold.[11] Inside the plants where they worked, the conditions and the methods of cleaning the broilers were crude. The historian William H. Williams described it this way,

> . . . the birds were hung on an overhead conveyer line after they were killed. Next, they were semi-scaled, rough-picked with mechanical pickers, waxed, and then the wax was removed to take away the finer feathers. Finally the broilers' skins were singed for any remaining feathers and residual food was removed from their crops.[12]

The typical work in the processing plants during the 1940s was crude. The workers dressed in white aprons and lined up around a u-shaped shackle hanger. Each worker manually checked the chickens as they came through the conveyor belt. Then other workers dipped the chickens in hot and boiling water to remove their feathers. The chickens were dipped in a particular way to avoid excessive peeling of the skin. Another group removed the intestines. At the other end of the conveyor belt, there were workers who weighed and then cut up the chickens before they were packaged in customized boxes and then stored in a separate temperature controlled room. The set room temperature was to ensure compliance with USDA meat storage regulations that required meats for public consumption to be stored at the freezing mark for public health reasons.[13] Throughout these different phases, many workers typically did not wear protective gear, which left them vulnerable to severe accidents. They could not wear the protective gear because it was expensive in comparison to their earnings.[14]

During the early phase of commercialized poultry production, virtually no woman worked as a chicken catcher. Black men mostly performed this task.[15] Catching chickens at this time was a violent and unpredictable exercise. One catcher stated that:

> During the 1930s and 1940s, the chickens were caught in the same way that they are caught today. One difference was that in those days when the chickens got bigger, the farm owners allowed them to roam about in open but fenced farms and they were hard to catch. At least the air was not as bad in the open farms. But most of the catching was done inside where the catchers went in, grabbed the chickens, sometimes in sets of four on each hand and then threw them into specially built crates. They repeated this motion many times and worked for long hours. The chicken houses were hot during the summertime. You had to go outside for some fresh air because of the smell. They had fans but only to keep the chickens cool. The catchers had to catch a lot of chickens.[16]

Men typically caught the chickens during the early hours of the morning to minimize injuries that they would have sustained at hours when the chickens were fully awake, restless, and aware of intruders. Men caught the chickens not because women could not do it; they could and were capable of doing so. But men dominated because the women worked in the processing plants.[17] However, processing plant work was not exclusively reserved for the women. There were men who also worked in the plants in addition to working as chicken catchers to supplement their low wages.

Chicken catchers typically caught between "6 to 16 metric tons of broilers for a standard 1,000-broilers caught per hour" in an eight-hour shift.[18] On the Lower Shore counties, catchers were paid about ninety-three cents per thousand chickens caught in the 1940s.[19] At this rate, catchers had to catch tens of thousands of chickens to save enough money to purchase their own homes. These catchers who were mainly African Americans were central to the early success of commercial poultry production on the Lower Maryland Eastern Shore. The commercial producers relied on them to move the chickens once they reached the broiler stage to the processing plants. If the catchers did not get the chickens to the processing plants at the right time, the chickens either died or were vulnerable to diseases. In both cases, it meant huge loses to the producers.

Commercial poultry production on the Lower Maryland Eastern Shore occurred at a time when black veterans returned home from World War II and the Korean War. Poultry processing plant jobs in particular were available for these black returnees as other economic avenues were restricted to them because many companies would not hire blacks.[20] Many of the black soldiers who returned home from the battlefields in the 1940s and 1950s needed to be reabsorbed into the local economies as well as into their previous occupation as farm workers. There were federally sponsored programs to assist black farmers to preserve black-owned farms; however, many black farmers lost their farms because they did not receive the relief assistance that was promised to them and were misled to believe that they were not entitled to such relief.[21]

African American involvement in commercial poultry production occurred in an environment of intense racial segregation and lively protests and demonstrations led and orchestrated by activists such as Gloria Richardson and Enez Stafford Grubb.[22] Other activists from the Student Nonviolent Coordinating Committee (SNCC), notable among them H. Rap Brown and the Freedom Riders, came down to the Shore in a show of solidarity for the local black strugglers. Their struggles were made more difficult by the activities of hate groups such as the Ku Klux Klan, which subjected them to acts of intimidation and harassment.[23]

By the 1960s, commercial poultry production on the Lower Eastern Shore faced stiff competition from southern poultry producers. However, the industry (on the Delmarva Peninsula) retained its rank as the second largest poultry producing region in the nation behind Georgia, which had a unionized commercial poultry workforce. Part of the reason for this stiff competition and Georgia's success was in the differences of wages paid to workers in unionized and non-unionized poultry processing plants. Commercial poultry production was non-unionized on the Lower Eastern Shore. Workers did not engage in union activities because of possible reprisals if they went against the companies and their supervisors.[24] In unionized plants, workers were paid about $1.40 per hour on the average while in non-unionized plants they were paid about $1 per hour.[25] Another factor that explains why commercial poultry production faced competition was the fact that corn and soybeans, two crops that were fed to the chickens, had to be imported from other places.[26]

In spite of the harsh working conditions on the Lower Maryland Eastern Shore, poultry workers contributed to Somerset County's profitable sales of $13,358,133 worth of poultry and poultry products in 1964, while Wicomico County had the most sales at $23,552,905, and was closely followed by Worcester County at $18,659,817.[27] In the entire Delmarva Peninsula, for example, commercial poultry accounted for $200 million in sales in 1967 and the money boosted the local economy.[28] These profitable sales were necessitated by the presence of poultry processing plants.

By the late 1960s, there were five poultry processing plants on the Lower Eastern Shore—two in Wicomico County (Perdue Plant in Salisbury, and Armour and Company in Wango); and three in Worcester County (Ralston Purina Company in Berlin, Showell Poultry in Showell, and Maryland Chicken Processors in Snow Hill).[29] There were no processing plants in Somerset County during the same period. The locations of these processing plants primed the Lower Eastern Shore for its preeminent role in commercial poultry production in Maryland. The labor force in these plants contributed to the economy of the Lower Eastern Shore in particular and Maryland's economy in general.

In the 1970s, there was a relative improvement in the working conditions in commercial poultry production when compared to the previous years. Prior to the 1970s, there was no federal or state agency in Maryland that was dedicated solely to monitoring workplace safety in the poultry processing plants or even in the poultry houses. The U.S. Department of Labor had oversight responsibility for ensuring workplace safety; however, the establishment of OSHA and MOSH in the 1970s led to the enforcement of workplace safety regulations. It was not uncommon before the 1970s, for line workers with finger cuts suffered while operating the machines to be sent back to the line after band aid and minimum first aid had been applied to stop the bleeding.[30]

The workers' accounts painted a picture of hazardous working conditions. However, as a result of OSHA and MOSH regulations, poultry companies were required to send injured workers to the hospital for full medical care and consultation or have a qualified medical professional onsite to properly treat workplace related injuries. Despite workplace safety issues, in the 1970s, poultry workers on the farms and in the processing plants remained productive.

Commercial poultry production in the United States in general during the 1970s and 1980s was a profitable venture. The industry experienced rapid growth due to a number of factors including advances in poultry breeding, nutrition, disease control, better housing designs, automated feeding machines, and temperature control equipment.[31] These factors enabled commercial poultry producers to earn millions of dollars in profits. Perhaps the most significant of the factors was the labor of the poultry house and processing plant workers who were responsible for ensuring that the reasons for the rapid growth translated to financial gains for the companies. However, in the case of Perdue Farms on the Lower Maryland Eastern Shore, and despite its contention of losing ground in the volume of poultry sold, in 1984 it grossed $840 million in sales.[32] This profit margin would not have been possible without the frontline workers. The profits did not translate to higher wages for the workers.

In 1987, (three years after the company grossed $840 million), commercial poultry workers, industry wide, earned on the average $5.87 an hour, hardly a reflection of a prosperous industry.[33] Despite the relative low wage that the workers earned, commercial poultry continued its upward climb in the 1980s.[34] For the industry as a whole, the 1980s saw an increase both in chicken consumption and sales. The medical profession was instrumental in the increase in poultry sales. For example, in the mid 1980s medical professionals contended that chicken consumption carried many health benefits over red meat.[35]

Commercial poultry producers capitalized on this finding. A series of sustained advertisements helped expand poultry's popularity with consumers. These advertisements lured consumers away from beef consumption, known for its high fat content. One example featured Frank Perdue proudly proclaiming the superiority of poultry meat while he taunted beef producers with the famous "who cares where the beef is" slogan.[36] The success of this chicken advertisement campaign catapulted Perdue to the top-tier of national food producers and made him the force behind commercial poultry production on the Lower Maryland Eastern Shore.[37] In 1987, poultry and poultry products constituted 57.9% of the total value of poultry and livestock sold in Maryland.[38] Despite these successes, there remained a need to improve the working conditions in the commercial poultry production facilities on the Lower Maryland Eastern Shore.

By the 1990s, commercial poultry production on the Lower Maryland Eastern Shore had an established labor force. During the 1990s, commercial poultry producers were criticized by advocacy groups for poor wages and unsafe working environment. In 1999, Leona Trotter and other processing plant workers filed a lawsuit against Perdue Farms in which the workers argued that they were not paid for the time they spent donning and doffing safety and sanitary equipment before and after working on the processing line. Perdue argued that they were not entitled to such pay because they were not officially on the time clock, since the workers' time began once they were at the processing line.[39] In a separate but similar case, processing line workers at a Tyson processing plant at Albertville, Alabama, filed a class-action complaint in 1999 alleging that they were required to work off the clock in violation of the Fair Labor Standards Act.[40]

During the 1990s and despite increased demand, consumption, and profits, a series of investigative reports by the media detailing consistent patterns of abuse and neglect of poultry workers attracted a closer scrutiny of the industry. According to a report from a Department of Labor probe in 1997, "60 percent of poultry companies were in violation of basic wage and hour laws."[41] The failure to pay workers the basic wage was not because the workers, many of them black, did not satisfactorily perform their jobs. Rather, they were not paid in part because the workers did not understand the labor laws and their complexities. Another consideration was the weak enforcement of labor wage laws in the chicken plants. The poultry companies provided a bulk of the local employment and many of the workers had limited skills and were at the mercy of the companies.[42]

Poultry workers were reluctant to speak out for fear of losing their jobs. There was a sense of a hostile work environment that permeated commercial poultry production.[43] Plant supervisors demanded loyalty from the workers and worked assiduously to compel the workers to tow the line. Workers who did not tow the line were sometimes assigned to work in more difficult places such as chiller rooms where room temperatures were colder than in other areas. In an effort to get the workers to tow the line, they were discouraged from speaking to outsiders about the inner workings of the plants. Perhaps a lack of education and therefore lack of knowledge of the laws may have been one important reason why many commercial poultry workers felt trapped with no other place to go. The workers tolerated the poor working conditions because they had not experienced the opportunity to work in a different occupation besides a poultry processing plant. The workers' lack of familiarity with laws against workplace abuse contributed to the companies' ability to limit and control information to the workers, lest they became emboldened to challenge some of the companies' workplace practices.[44] The less the workers knew the better for management.

There were several instances of worker injury, but many of the injuries were not officially reported. The fewer the reported number of injuries at the plants, the better for the supervisors because the reported injuries were part of their supervisory evaluative measures.[45] There was a well-founded reason for the supervisors to control the flow of information concerning workplace practices. One case in point involved Patrick Harmon who worked in the Berlin processing plant. He was injured on the job, but was told by a nurse and his supervisor that his injury was minor and was sent back to the line. Harmon's injury was severe to the extent that it affected his mobility and required surgery.[46] He should have been entitled to work-place protections such as workman's compensation, which was denied him because supervisors kept helpful information away from him and instead gave him wrong information.[47]

Many of the workers interviewed in this study stated that workers were required to pick up the workload of those who called out sick without any additional pay for the additional work. Their individual quotas were increased without any adjudication of whether their workload was commensurate with what they were paid. Some of the workers accepted such practices and arrangements as the norm. This practice only began to change in the 1980s after union representatives and organizers began to interact with the workers, usually after work, because they could not organize on company time and property. In their interactions, the organizers gleaned information and evidence that were subsequently used in contract negotiations between workers and management. Since the workers had no one to formally speak out on their behalf, UFCW stepped in and joined forces with other interested groups and began grassroots organizing to highlight the problems in the workplace, as well as issues regarding fair wages and compensation.[48]

Before the late 1990s, black workers did not wholeheartedly embrace the UFCW. There was the suspicion on their part as to whether a predominantly white labor organization really understood the conditions and struggles of black poultry workers. However, these concerns were eventually put to rest after the union worked with other advocacy groups in 1999 to file lawsuits against Perdue Farms and Tyson Foods for back pay/overtime violations.[49] Meanwhile, the number of Latino workers in the poultry industry in the Lower Maryland Shore grew and the UFCW was able to attract the membership of many black and Latino workers.

The presence of the union along with its activities emboldened the workers to agitate against the practices of under-payment and wage withholding by some of the poultry companies. In one celebrated case, *Harmon* v. *Tyson*, Patrick Harmon, an African American chicken catcher, along with 95 other catchers at Tyson's Berlin, Maryland, and Temperanceville, Virginia processing plants, sued Tyson in July 2000 for failure to pay them for overtime work dating back to 1997. Tyson argued in its defense that the catchers were not company employees and therefore, were not entitled to overtime pay.[50]

The case against Tyson Foods came about after Harmon, who had worked for several poultry companies including Perdue Farms was injured on the job. Prior to his injury, he had worked long hours, beyond his normal work hours as a chicken catcher. He decided to sue after he was fired for being out with his injury.[51] The catchers were contracted to work for a certain number of hours. If they exceeded the contracted hours, the Fair Labor and Standards Act required that they be paid for the additional time worked. The workers did not receive the overtime pay. With assistance from the Public Justice Center (PJC), UFCW, and DPJA, Harmon's case became a class-action lawsuit and in 2001, Tyson agreed to pay $2.4 million to settle the lawsuit.[52]

After a series of similar lawsuits in other parts of the country against major poultry companies over unfair labor practices, their contention that chicken catchers were non-company employees and denial of labor wages were no longer persuasive and could not convince the courts. In 1999, a group of workers at IBP, Inc., a large meat processor in Pasco, Washington filed a class action lawsuit claiming unpaid wages.[53] In the same year, a group of poultry plant workers in Portland, Maine sued their employer Barber Foods Inc., "to recover compensation for alleged unrecorded work covered by the Fair Labor Standards Act of 1938."[54] These cases and the rulings by the courts in favor of the workers corroborated the allegations of unfair wages leveled by former and current processing plant workers.

In a sense, the legal victory by Harmon and the other plaintiffs underscored the intimidation that poultry workers faced. The victory gave them legal empowerment. In the case of Harmon, the nature of his work was similar to a freelance who maintained different jobs because he was not paid enough by one company to survive. Harmon had another job, as was the norm for many poultry workers who had difficulty making ends meet with such low wage jobs as working in the plant, or catching chickens. Harmon's suit against Tyson had a domino effect as labor rights lawyers, many of whom worked *pro bono* on behalf of the workers, brought cases of these sorts.[55]

It would seem that as the largest poultry producer in the United States, Tyson could pay overtime to the workers with minimal impact on their profit margin. As one of the plaintiffs in the case appropriately stated, "I hope the company realizes they cannot treat us any old way, and that we are an important part of the process."[56] In yet another noted case in February 2000, catchers sued Perdue and won nearly $2 million in "back wages."[57] In each of these cases, the companies were forced to reassess their treatment of chicken catchers.[58]

These catchers were important to commercial poultry production. The processing plants cannot process the broilers until they have been caught and sent to the plants. The growers relied on the catchers. Tyson's argument was the classic cover-your-tracks syndrome where companies engaged in abuse of workers but used loopholes in contracts or laws to absolve themselves from civil and criminal liability. The company's suggestion that the catchers were independent was a weak contention. As it was, the catchers could not have caught the chickens in a vacuum, since they were part of the overall operations of the big companies. Even many of the growers, who would better be described as independent contractors, were dissatisfied because of the arrangement that kept them bound to the poultry integrators, whom the growers saw as a monopoly.[59]

As a result of the court rulings, the major companies on the Lower Shore—Mountaire, Perdue, and Tyson—reworked their benefit packages to include retirement pension plans and medical insurance, both of which had been lacking or flawed in the previous two decades. Due to the pressure and spotlight on the working conditions of poultry workers brought about by Harmon's lawsuit, chicken catchers became company employees and eligible for benefit packages including health insurance, retirement pay, paid vacations and holidays.[60]

These lawsuits and the media spotlight made the companies to confront the moral dilemma of making huge profits while paying sub-standard wages. Without the lawsuits, labor laws may not have been enforced, workplace safety might have gone unchecked, and working conditions might have remained appalling. Despite the poor working conditions, by 1999, the UFCW represented only 40 percent of the processing line workers on the Lower Shore.[61] The level of participation in union activities was relatively low mainly because the poultry companies were adept at playing down the relevance of the union. They used a combination of tactics such as occasional one-time bonus payments to workers and company-sponsored activities to make workers think and believe that they do not need the union. They used these tactics to give workers the impression that the companies cared about them and that union involvement was not necessary.[62]

Commercial poultry production lured many Latino workers to the Lower Maryland Eastern Shore. Soon after Frank Perdue took to the airwaves in the 1970s and portrayed the industry in positive terms and as a friendly place to work, Perdue Farms along with other commercial poultry companies went on an aggressive recruitment of Latino workers. An end result was the growth of Latino workers mainly because of the availability of poultry and other service related jobs.[63] Most of them had low educational level and were not fluent in English. For those who became part of the poultry labor underclass, much education was not required because of the non-complexity of catching chickens. Also, much education was not required for working on the lines inside the processing plants. In essence, no specialized educational or English language skills were required of the Latino workers. The main requirement was that they be physically able-bodied to engage in repetitive motions that were the standard in the plants.

Most of the Latino immigrant workers were unskilled and were drawn to poultry processing plant work because of the immediate cash payments.

Their work brought them "good wages" and enabled them to earn U.S. currency, which exchanged for larger sums of money in their respective countries' currencies.[64] Prior to the Latino migration, poultry work was once a "cherished" occupation on the Shore.[65] For many Latinos on the Shore, working in the processing plants or catching chickens was better than not working at all.[66] The Latino workers did not need specialized training to perform this sort of low wage work. Since they faced a language barrier (most of them did not speak English), it worked to their advantage that they worked in a field that required lesser communication with other line workers. They welcomed the opportunity to work for a livelihood in a place that was culturally different but latinized.[67]

COMMERCIAL POULTRY PRODUCTION AND LABOR ACTIVISM

Labor activism in commercial poultry on the Lower Maryland Eastern Shore dated back to the 1950s with the establishment of the Eastern Shore Growers Poultry Exchange in Selbyville, Delaware in 1952.[68] This Exchange was not formally identified as a labor union. It was more of a growers' cooperative, but the Exchange was an outlet for independent poultry growers. To an extent, the Exchange was successful because it enabled growers to sell their poultry directly to local and outside companies. By the time the Exchange closed in 1969, growers wielded some amount of power in terms of determining the amount of flock that they raised, and the profits that they made from their poultry. Had the growers not participated in this Exchange, it is plausible that they may not have benefited from the economic opportunities that the Exchange provided.

In the 1970s and 1980s, labor activism in commercial poultry production was uneventful because of anti-unionism by poultry companies. The ownership of poultry companies were in the hands of fiercely local and independent-minded farmers who were risk takers and valued the classic American agrarian ethos and resented attempts to impose outside influences.[69] The consolidation and acquisition of smaller poultry operations on the Lower Eastern Shore also strengthened the power and influence of large producers such as Frank Perdue who preferred his plants to be non-unionized.[70] The ownership and leadership structures of the companies denied access to outsiders about the inner workings of their operations. This denial strategy limited labor union activism particularly on the Lower Maryland Eastern Shore.

However, the tempo and momentum of labor activism intensified from the 1990s. The era of the 1990s was different in the sense that it was the onset of vocal advocacy and activism that focused more on the workers in the poultry houses and processing plants. As other occupational choices brought on by the diversification of the local workforce became available, as well as the workers' recognition of their role in keeping the companies afloat, more scrutiny was brought to bear on the industry by labor groups and other workers' rights groups. Such scrutiny made it difficult for commercial poultry producers to ignore the working conditions in the poultry houses and processing plants. It was the age of labor radicalism in the commercial poultry industry. Leading the vocal criticism against commercial poultry production and the unfair working conditions in the 1990s was the Delmarva Poultry Justice Alliance (DPJA). The Alliance was a forceful advocate for safe working environment, fair treatment of animals, safety of all poultry workers, fair wages for workers, fair poultry grower contracts, and environmental protection of the Chesapeake Bay. The Alliance was a staunch supporter of the poultry labor underclass, and it was instrumental in promoting collective struggle of African American and Latino poultry workers, as well as white growers. This class collective struggle rallied commercial poultry workers on the Lower Eastern Shore together to agitate for better working conditions in the processing plants.

The addition of Latinos into the commercial poultry labor underclass is instructive. As more options for employment became available for blacks, many of them left their positions as chicken catchers, eviscerators, gut removers and poultry meat packers. Latinos, mostly Guatemalans and Salvadorians and a small number of Laosians emerged as the new immigrant group of the poultry working class.[71] They constituted a significant number of the poultry labor pool for the industry on the Lower Maryland Eastern Shore. The number of Latino workers steadily increased in the areas of catching, transporting, slaughtering, and processing broilers.[72] The increased number of Latino workers in commercial poultry production was not only confined to the Lower Maryland Eastern Shore. According to the U.S. Bureau of Labor Statistics, 25 percent of the poultry workforce in the United States in 1994 was Latino, while 46 percent was white and 25 percent was African American.[73] By the late 1990s, Latinos surpassed both African Americans and whites as the dominant workforce in commercial poultry production.[74]

However, a small number of Asians of Vietnamese backgrounds managed to secure a new role as poultry growers.[75] These Vietnamese growers were not new to raising commercial poultry. According to the *CIA World Fact Book*, in the 1980s and 1990s, 63 percent of the Vietnamese labor force was agriculturally based and 21.8 percent of its Gross Domestic Product was in agriculture. Poultry constituted a major part of its agricultural sector.[76]

The Vietnamese and other Asians' participation as poultry growers was aided by a program called "Loans for Beginning Farmers and Ranchers" initiated by the United States Department of Agriculture in the 1980s. It was intended to allow interested persons to gain access to capital to run a farm operation such as poultry growing at an interest rate lower than what commercial lenders charged. The Asians were one of the groups identified by the USDA as socially disadvantaged (SDA) and hence eligible for this program.[77] Subsequently, this loan necessitated their receiving growers' contracts. Local white growers were disenchanted by the terms and conditions of the contracts that Asian growers received from the large-scale commercial poultry companies. The white growers argued that the Asian growers accepted lesser terms because they were not vested in the communities and could leave at any time if they did not make much profit. The local growers, on their part, were vested in their communities and stood to lose more of their investments including their homes, if they defaulted on the loans.[78]

Although African Americans were part of the targeted SDA groups, they were less successful in obtaining these farm loans in part because commercial poultry companies were still reluctant to extend growers' contracts to African Americans. Even though they were vested in their communities, African Americans still had to deal with issues of established credit history and worthiness. In fact, it was the practice that before poultry growers received the loans; they had to provide evidence that they had a contract to grow chickens from one of the companies.[79] Consequently, the inability of blacks to obtain these loans affected their status and position in the commercial poultry production on the Lower Maryland Eastern Shore in the sense that only a negligible number were contract growers.

The presence of Asians on the Lower Maryland Eastern Shore began in the early 1990s. Under a program sponsored by the U.S. government to fill unskilled jobs in the U.S., willing Korean and immigrants from other nations received visas to travel to and work in the United States.[80] They were granted permanent resident visas on the condition that they will work in the poultry plants. As a condition for the permanent status, the poultry companies must have shown a need for them and that the companies could not find American born workers to fill such job vacancies. The companies also must have advertised widely for workers to come to the United States with the agreement that they would fill the needed jobs.[81]

This program, which was jointly administered by the U.S. Labor Department and the defunct Immigration and Naturalization Service (INS), was an important tool for poultry companies; because, it allowed them to replace a dwindling local labor force many of who had better employment prospects in other sectors of the economy. Local workers were no longer drawn to processing plant work as before, and the arrival of Koreans and other Asians was a welcomed relief for commercial poultry industries. Their arrivals at the plants were not without exploitation. These Koreans, before leaving their home countries, paid large sums of money to intermediaries who procured them for some of the poultry companies.[82] The Korean immigrants paid large sums of money because the long-term benefit of holding a U.S. permanent residency status compensated for their hardship and obligation to fulfill the terms of their visa agreement. Under normal circumstances, many immigrants who came to the U.S. to work usually accepted jobs that locals refused for various reasons. After working in these low paid jobs, many of them made the choice to either continue working in their initial places of employment, or found other higher paying jobs with better conditions and other incentives. With the Korean immigrants and others in similar conditions, they recognized the temporary nature of their arrangements as many of them left poultry work and went into other higher paying occupations, after they fulfilled their contractual obligations.

The Korean immigrants were led to believe that they were moving to a life of freedom without realizing that they had signed on to sustained wages exploitation. Many Asian workers were happy to live and work in the United States. Many of the Korean workers paid thousands of dollars to come to the United States even with no firm guarantee of "decent employment."[83] Upon arrival on the Lower Maryland Eastern Shore, the Asian workers became part of the poultry labor underclass and encountered multiple challenges including language barriers and cultural differences. They built and maintained a small and tightly-knit community and engaged in language training skills development initiatives, while attempting to overcome the barriers that they faced.[84]

The emergence of labor radicalism in the 1990s that involved blacks, Latinos, Asians, and Caribbean poultry workers coincided with a rise in direct activism by concerned clergy, environmentalists, and labor unionists through the DPJA. The Alliance argued against the conditions in which the poultry workers labored for the companies as morally and socially unjust and used direct tactics in terms of physical demonstrations and protests on the properties of Perdue Farms and Tyson Foods to make their case. The Alliance's quest for social justice hastened and intensified poultry labor radicalism in the 1990s.

Activists and demonstrators were arrested several times for trespassing on the properties of commercial poultry facilities that belonged to Perdue Farms and Tyson Foods on the Maryland and Delaware sides of the Delmarva Peninsula. The Rev. Lewis was the main organizer of several direct actions mainly because he tapped into his connection with members of the press. Every demonstration that he organized always attracted the attention of the media. He was interviewed by Mike Wallace of the CBS News program, *60 Minutes,* (19 December 1999). While the workers had agency in terms of articulating their concerns about working conditions in the poultry houses and processing plants, the typically well-orchestrated and publicized demonstrations drew large media coverage, which worked to the advantage of the workers and gave them a larger platform. Had such action been left to the workers' agency, perhaps they could have obtained a different outcome. These advocates, along with the workers and the UFCW viewed wages and working conditions as gross miscarriages of social and economic justice. Consequently, these issues became the central focus of commercial poultry radicalism on the Lower Maryland Eastern Shore.

He incessantly criticized commercial poultry practices of vertical integration, grower contracts' termination without adequate notice, and environmental degradation. A retired priest from the Episcopal Diocese of Delaware, he spearheaded the creation of the Delmarva Poultry Justice Alliance (DPJA) in 1997.[85] As an activist, the Reverend Lewis used his moral authority to rally exploited growers, catchers, unions and other interested parties to the cause of fairer contracts, better conditions and higher wages for poultry workers. He brought poultry growers, catchers, and line workers to work together against "exploitation by Big Chicken."[86]

A notable event undertaken by the DPJA and UFCW occurred in 1999. On October 8 1999, a worker named Charles Shepperd, who worked for Tyson Foods at its Berlin, Maryland processing plant, fell into a chiller room where chickens were refrigerated after they had been processed. Shepperd's fall crushed his skull when the machine that operated the chiller room was alleged to have been "either accidentally or intentionally" activated by one of the workers.[87] This accident, along with five other deaths that involved employees of Tyson Foods and their subcontractors, was investigated by OSHA. The investigation concluded that the deaths did not all result from injuries sustained on the poultry processing assembly line, and that some of the deaths were due to natural causes.[88] The dangers of working in these plants prompted an official of OSHA to warn in 1998 that "too many plants were focusing on production at the expense of employee health."[89] Although commercial poultry producers showed a willingness to improve the working conditions and worker safety in the 1990s by devoting more time to employee safety training sessions, the death of Shepperd in 1999 showed that working in the processing plants remained a major safety concern.

The Tyson Berlin plant accident and death, received much publicity in part because of the aggressive activism of both the DPJA and UFCW. Although this death received widespread publicity and painted the commercial poultry industry as prone to fatalities, according to the Maryland Occupational Safety and Health (MOSH), there were only three fatalities in poultry slaughtering and processing in Maryland from 1992–2001.[90] The depiction of commercial poultry production as prone to workplace fatalities was in line with efforts by the DPJA and UFCW to capitalize on one, or a few negative incidents that strengthened their attempts to pressure the companies, to improve the working conditions in the poultry houses and processing plants. Critics used worksite-related accidents and deaths including findings by the United States Government Accountability Office (GAO) that, "meat and poultry workers are injured in a variety of ways, and their injury and illness, though declining, remain among the highest of any industry," to further contend that commercial poultry workers deserved fairer and safer working conditions.[91] The companies had primary responsibility to comply with OSHA standards to ensure their workers' safety. As a result of the death of Shepperd, Tyson Foods and other commercial poultry producers on the Lower Maryland Eastern Shore re-examined their safety management measures.[92]

The labor radicalism of the 1990s brought both the Latino and black poultry workers to collaborative resistance. These communities were united in a wage struggle against "Big Chicken" on the Lower Maryland Eastern Shore. In many ways, the black and Latino workers took the same work ethic of consistency, commitment, perseverance and determination that they showed in their factory jobs, to their quest for change by participating in demonstrations, protests, vigils, and meetings. Although they did not live in the same community or socialized at the same places, their collaborative effort created a sense of collective communalism that inspired their agitation for better working conditions. Their joint efforts and those of other advocacy groups helped to bring about fairer wages and worker safety.[93]

As the local poultry workforce became predominantly Latino by the 1990s, there were bi-lingual programs to address the language gap since many of the Latino workers who arrived on the Shore did not speak English. Part of the reason for such programs was not necessarily to help the companies better understand the workers or for the workers to know all about the companies. Rather, the purpose of these bi-lingual programs was to educate the workers about their rights under U.S labor laws and to demand fairness in the workplace. The hope was that the new workers would speak up and speak out when they were asked to work long hours without pay. At the same time, the new workers, by demanding to be paid standard wages, refused the low wages that some of the companies paid their Latino workers, which was less than the normal and legal wage.[94]

To further discourage workers from engaging in union activities, Perdue Farms sought to mechanize its chicken catching operation in the 1990s. This was a subtle attempt to threaten the means of livelihoods of catchers who agitated for improved working conditions in the chicken houses. With the threat of mechanized chicken catching removed, because it was incompatible with maintaining the quality of poultry meat, the catchers opted for union representation to guard against further attempts by poultry companies to threaten their jobs.[95]

The mechanical system and approach to chicken catching, which required a conveyor belt to scoop the chickens into a recessed belt for onward herding into a ventilated cage had been attempted in Georgia in the 1970s without much success.[96] Poultry mechanization did not experience much commercial success on the Lower Eastern Shore because it could not guarantee the preservation of the quality of the chickens in the processing plants. The chickens that went through the mechanized process showed signs of bruises and did not meet the quality standards of the producers. Proponents of mechanized catching thought it to be a cost-saving measure, as it was an attempt to minimize the cost of production and maximize profit.

The machines were another attempt to quiet the vocalism of the union and the advocacy groups. These groups spoke out on behalf of the catchers and were concerned about loss of jobs and benefits. The machines simply shifted worker exploitation in a different direction. In this case, after working for the companies and being a crucial factor in the profit building, the catchers, were simply denied the rewards for their labor as Perdue Farms attempted to replace them with machines.[97] The companies might have considered the machine option to solidify their contention that the catchers could not claim overtime from Tyson. It was their argument that only company employees were eligible for overtime pay.

CONCLUSION

Despite various working conditions, commercial poultry continued its expansion during the large-scale commercialization of poultry. The poultry companies on the Lower Maryland Eastern Shore benefited from the hard labor of the working and underclass. The large volumes of poultry that were produced and sold on the Lower Shore demonstrated the critical role of the workers. In particular, black workers played an important role in the economic viability of the region. First of all, poultry was central to the regional economy, and second, black workers constituted the majority of employees at commercial poultry processing plants.

The arrival of Latino workers on the Lower Shore in the 1970s expanded the poultry underclass due to their low level of education. Their underclass status became more pronounced in the 1990s. They dominated the poultry processing assembly line as many black poultry workers gradually left to work in other sectors of the local economy such as tourism. The increased involvement and activism of the DPJA and the UCFW in the 1990s brought the Latino and African American communities together in a struggle to fight the domination of "Big Chicken." The DPJA and UCFW played important roles in raising the consuming public's awareness of the working conditions in the poultry houses and processing plants. It was a period that juxtaposed callous capitalism with workers' rights and fairness.

Poultry workers on the Lower Maryland Eastern Shore were largely responsible for the large profits that were accrued by commercial poultry producers such as Perdue Farms and Tyson Foods since the early and large-scale phases of the commercialization of poultry production.[98] In the 1980s, Perdue Farms, for example, grossed over $800 million in sales in just one year.[99] Then in the 1990s, the company's sales exceeded $1 billion. The workers' wages did not rise in proportion to the large profits that the companies accrued from sales and consumption of poultry meat. Such disparity gave ammunition to critics of the industry and seemed to have strengthened their arguments about unfair wages and exploitation of workers. The disparity also empowered advocacy groups such DPJA and UCFW who argued for better working conditions and compensation for the rank and file workers of the industry.

NOTES

1. See *Maryland Agricultural Statistics Reports* (Washington, D.C: United States Department of Agriculture; various years) showed a steady increase in the volume of chickens raised and used for commercial purposes statewide and particularly on the Lower Eastern Shore from the 1950s through the 1990s.

2. Brendan Sexton, "The Working Class Experience," *The American Economic Review*, Volume 62, No. 1/2 (1972), 149–153.

3. Raymond P. Smith Sr., interview by author, Berlin, Maryland.

4. Human Rights Watch, "Blood, Sweat, and Fear: Workers' Rights in U.S. Meat and Poultry Plants," Available at http://www.hrw.org.

5. The Occupational Safety and Health Act of 1970 was enacted by Congress with a mission "to send every worker home whole and healthy everyday." This Act established the Occupational Safety and Health Administration. The implication from this agency's mission indicated the occupational hazards that many commercial poultry workers faced on the job. See http://www.osha.gov. Several processing plant interviewees for this study stated that accidents were common within the plants.

6. See http://www.dllr.state.md.us/labor/mosh for more information on MOSH.

7. Several workers interviewed for this study narrated their personal experiences of sustaining injuries on the assembly line or in the chicken houses and having band aids applied to their injuries and being sent back to work. These injuries were not

catalogued or reported as required by OSHA and MOSH regulations. In the end, inspection officials did not have any reason to cite or impose sanctions on the companies for workplace safety violations, even though they occurred, because company officials did not report such injuries as were reported to them by the line or poultry house workers.

8. See "Estimated Number of Workers in April 1939, Subject to Provisions of the Fair Labor Standards Act" (Washington, D.C.: U.S. Bureau of Labor Statistics, 1939) [U.S. Department of Labor Library, Washington, D.C.]: 6 and 10. Specific numbers for poultry workers in Maryland in general and the three counties in particular were not available.

9. According to the U.S. Department of Labor, Wage and Hour Division, and under the Fair Labor Standards Acts of 1938, Title 29, Part 780 (8) and the *Interpretative Bulletin*, poultry workers were classified into two categories—those involved in "raising" and those involved in "slaughtering" poultry. The workers in each category received different pays. In 1939, the federal minimum wage was 25 cents an hour. Those involved with raising poultry received the minimum wage while those engaged in slaughtering received between 35 and 40 cents an hour. Those engaged in the slaughtering were exempt from the overtime wage law. See *Interpretive Bulletin*, No. 14, 1938, p.15. From this earlier stage then, even the federal government recognized the nature of the work involved in the processing plants by enacting pay differentials within the poultry labor force.

10. William H. Williams, *Delmarva Chicken Industry. . .* , 31.

11. *Campbell's Salisbury Plant Memories, 1946–1993*, (Salisbury: Campbell Soup Company, n.d.) [brochure, Wicomico Public library].

12. Williams, 33. Cleaning methods in the processing plants have improved since the 1930s, 1940s, and beyond. New workplace regulations instituted by the Department of Labor have led to improved sanitation as well as working conditions for commercial poultry workers. See http://www.osha.gov.

13. Spotwood Jackson, interview by author, Baltimore, Maryland. See also, http://www.fsis.usda.gov.

14. Spotwood Jackson interview. For visual images of workers inside the processing plants in the 1940s, see Williams, 41.

15. Spotwood Jackson interview.

16. Raymond P. Smith Sr., interview by author.

17. The Reverend Thomas Sweatt, interview by author, Princess Anne, Maryland.

18. Shai Barbut, *Poultry Products Processing: An Industry Guide* (Boca Raton, Florida: CRC Press LLC, 2002), 61.

19. The Reverend Sweatt interview.

20. Enez Stafford Grubb, interview by author, Cambridge, Maryland.

21. Lu Ann Jones and Nancy Grey Osterud, "Breaking New Ground: Oral History and Agricultural History," *Journal of American History*, Volume 76 (September, 1989), 556.

22. See the following works: Sandra Millner, "Recasting Civil Rights Leadership: Gloria Richardson and the Cambridge Movement," *Journal of Black Studies*, 26 (July, 1996), 668–687; Edward K. Trever, "Gloria Richardson and the Cambridge

Civil Rights Movement, 1962–1964," (M.A. Thesis, Morgan State University, 1994); Sharon Harley, "'Chronicle of a Death Foretold': Gloria Richardson, the Cambridge Movement, and the Radical Black Activist Tradition," in Bettye Collier-Thomas and V.P. Franklin, eds., *Sisters in the Struggle: African American Women in the Civil Rights–Black Power Movement* (New York: New York University Press, 2001), 174–196; Arnette K. Brock, "Gloria Richardson and the Cambridge Movement," in Vicki L. Crawford, Jacqueline A. Rouse, and Barbara Woods, eds., *Women in the Civil Rights Movement: Trailblazers and Torchbearers, 1941–1965* (Bloomington: Indiana University Press, 1993), 121–144; Kathleen Thompson, "Gloria Richardson," in Darlene Clark Hines, Elsa Barkley Brown, and Rosalyn Terborg-Penn, eds., *Black Women in America: An Historical Encyclopedia*, Volume II (Brooklyn, NY: Carlson Publishing, 1993), 980–982, and Solomon Iyobosa Omo-Osagie II, "'Count Her In': Enez Stafford Grubb and the Building and Rebuilding of an African-American Community," *Southern Historian: A Journal of Southern History*, Volume XXIV (Spring, 2003), 40–49.

23. Enez Stafford Grubb interview.

24. See John Breuggemann and Cliff Brown, "The Decline of Industrial Unionism in the Meatpacking Industry . . . 1946–1987," *Work and Occupation*, (August 2003), 336 and passim.

25. John C. Schmidt, "On the Shore Chickens Lay Golden Eggs," *Salisbury Times*, 2 April 1961. [Maryland Room, Vertical Files, Wicomico Public Libraries, hereinafter referred to as Wicomico Library].

26. Ibid.

27. U.S. Bureau of the Census, Census of Agriculture, 1964. Statistics for the States and Counties, Volume 1, Part 23, Maryland, *Dairy Products Sold and Poultry and Poultry Products Sold*: 1964 and 1959, Table 12 (Washington, D.C.: U.S. Government Printing Office, 1967), 267.

28. Delmarva Poultry Industry, Inc., (DPI Archives) Georgetown, Delaware.

29. Mel Toadvine, "Two Processing Plants Add to Shore Economy," *Salisbury Times*, 7 April 1963. [Wicomico Library].

30. Patrick Harmon and Raymond Smith, interview by author.

31. F. Lasley and Allen Baker, "From Barnyard Flocks to Big Business," in *Farmline*, U.S. Department of Agriculture, Economic Research Service (Washington, D.C.: U.S. Government Printing Office, May 1984), 4 and passim.

32. John Bozman, "Delmarva's Competitive Edge Slipping," *Daily Times*, 28 October 1985.

33. Quoted from http://www.ufcw.org/press-room.

34. See Ziaul Z. Ahmed and Mark Sieling, "Two decades of productivity growth in poultry dressing and processing," *Monthly Labor Review Online*, Vol. 110, No. 4 (April 1987), 34.

35. Brice Stump, "Poultry Industry Experiences Big Boom," *Daily Times*, 10 August 1986.

36. Available at http://www.perdue.com/timeline.

37. Christian McAdams, "Frank Perdue is Chicken," *Esquire*, April 1975, p. 113–117.

38. U.S. Bureau of the Census, *Census of Agriculture, 1987. Maryland: State and County Data*, Volume 1, Part 20, *Value of Livestock and Poultry Sold*, 1987, Figure 7 (Washington, D.C.: U.S. Government Printing Office, 1989), Maryland-6.

39. See *Leona Trotter et al. v. Perdue Farms*, Civil Action No. 99-893. Available at http://www.ded.uscourts.gov. This case was filed in the United States District Court for the District of Delaware on 19 December 1999.

40. See *M. H. Fox, et al., v. Tyson Foods Inc.*, Civil Action No. CV-99-TMP-1612-M. Available at https://ecf.alnd.uscourts.gov. This case was filed in the United States District Court for the Northern District of Alabama, Middle Division on 22 June 1999.

41. http://www.ucfw.org/workplace. See also http://www.dol.gov.

42. The Rev. Jim Lewis, interview by author, Charleston, West Virginia.

43. Ibid.

44. Patrick Harmon and Alison Morton, interview by author.

45. Ibid.

46. Ibid.

47. The Rev. Jim Lewis, interview by author.

48. Patrick Harmon interview. Also, a separate interview with Bruce Drasal, Baltimore, Maryland.

49. See the following http://www.publicjusticecenter.org/poultryworkerscases, Sandy Smith, "Perdue Farm Settles, Tyson Foods Fights Donning and Doffing Disputes," *Occupational Hazards: The Authority on Occupational Safety, Health and Loss Prevention*, (10 May 2002), and *Leona Trotter et al. v. Perdue Farms,* Civil Action No. 99-893.

50. Joseph Cacchioli, "Chicken catchers' settlement with Tyson finalized," *Daily Times*, 3 November 2001. See also, Abosede George, "UFCW Joins Alliance Seeking Justice for Poultry Workers," http://www.laborresearch.org. Although most of the cases were either filed in the late 1990s, they were not adjudicated until 2000, 2001, and 2002. However, during the court battles in 2000 and 2001, Perdue Farms argued that chicken catchers were not employees of the company because they worked through sub-contractors who hired the catchers. But the U.S. District Court of Maryland disagreed and ruled that the catchers were company employees. See UFCW press release, "Perdue Chicken Catchers stand up for a Voice . . ." http://www.ufcw.org (4 May 2000). The issues in contention predated 2000, but the workers became more outspoken after the 1990s and their voices became louder through their collaborative activities with DPJA and the UFCW.

51. Patrick Harmon interview.

52. http://www.ufcw.org/perduefacts.

53. *Gabriel Alvarez et al. v. IBP, Inc.*, Nos. 03-1238; filed in the United States Court of Appeals for the Ninth Circuit. The Supreme Court ruled in favor of the workers.

54. *Abdela Tum et al. v. Barber Foods Inc.*, Nos. 04-66; filed in the United States Court of Appeals for the First Circuit. This case was heard and decided by the United States Supreme Court in 2005 in favor of the workers. See also Opinion of the Court: 546 U.S. ___ 2005, p. 15.

55. The Reverend Jim Lewis interview. He was instrumental in these *pro bono* cases. Large fees did not motivate the attorneys in these cases because the poultry workers simply did not have the money to pay lawyers. Instead, the *pro bono* work was motivated by the quest for social and environmental justice, two issues at the heart of Reverend Lewis' DPJA. He brought much publicity to the cause of the poultry workers who were seeking better wages and a safe working environment.

56. Cited in Joseph Cacchioli, "Chicken catchers' settlement with Tyson finalized," *Daily Times*, 3 November 2001.

57. Ibid.

58. The Reverend Jim Lewis interview.

59. Rick Thornton, "Poultry growers feel trapped in power struggle," *Daily Times*, 25 March 1992. See also, Christopher Thorne, "Farmers crying foul over Perdue," *Las Vegas Review Journal*, 25 December 1999.

60. Patrick Harmon interview. Also, interviews with the Rev. Jim Lewis and Carole Morison corroborated Harmon's assertion. Attempts made by the researcher to obtain further corroboration from Perdue Farms or Tyson Foods officials were unsuccessful.

61. Lena H. Sun and Gabriel Escobar, "On Chickens Front Line," *Washington Post*, 28 November 1999.

62. Patrick Harmon interview.

63. James Bock and Dail Willis, "Changing face of the Shore; Latinos: Lured by abundant jobs and good wages, Hispanic immigrants have become a sizeable community on the Delmarva Peninsula" *Baltimore Sun*, 13 October 1996.

64. Ibid. Bock's and Willis' characterization of "good wages" was really in comparison to the workers' previous status of substandard wages by American labor standards. When the currency exchange rate for the workers' home countries, many of which have traditionally weak currency rates, were converted, their wages translated to "good wages" in relation to their countries' standards because the workers typically repatriated their wages to their home countries and raised their standard of living as well as those of their families there.

65. I am using cherished here in the sense that while poultry work was labor intensive, at the same time it was stable and workers had a job to go to every day. Although the pay was small, they were guaranteed a regular income. So they cherished the job security that commercial poultry provided despite the challenging working conditions.

66. Lena H. Sun, and Gabriel Escobar, "Chicken Front Line . . ."

67. See David Griffith, "Hay Trabajo: Poultry Processing, Rural Industrialization, and the Latinization of the Low-Wage Labor," in Donald D. Stull, et al., *Any Way You Cut It: Meat Processing and Small-Town America* (Lawrence: University Press of Kansas, 1995), pp. 129–130.

68. William H. Williams, *Delmarva's Chicken Industry. . .* , 52–54.

69. Robert Bussel, "Taking on "Big Chicken": The Delmarva Poultry Justice Alliance," *Labor Studies Journal*, Volume 28, No. 2 (Summer 2003), 6. See also the following, George E. Pozzetta, *Unions and Immigrants: Organization and Struggle* (New York: Garland Publishing Inc., 1991); George E. Pozzetta, *The Work Experience:*

Labor, Class, and Immigrant Enterprise (New York: Garland Publishing Inc., 1991); Richard Alaniz, "Multiple Factors Influence Declining Union Membership," *Meat & Poultry*, (May 1998), 68; Charles Craypo, "Meatpacking: Industry Restructuring and Union Decline," in Paula B. Voos (ed*.), Contemporary Collective Bargaining in the Private Sector* (Madison: Industrial Relations Research Association, 1994), 63–96.

70. Ben A. Franklin, "Union Bashing 'Tough Man' to Organize Perdue Chicken Pluckers," *New York Times*, 22 December 1980.

71. Steve Striffler, "Inside a Poultry Processing Plant: An Ethnographic Portrait," *Labor History*, Volume 43, No. 3 (2002), 305–313.

72. "The Changing Face of Delmarva," *Rural Migration News* Volume 3, No.3 (July, 1997).

73. See "Workplace Safety and Health: Safety in the Meat and Poultry Industry, while Improving, Could Be Further Strengthened," A Report to the Ranking Minority Member, Committee on Health, Education, Labor, and Pensions," United States Senate [Government Accountability Office, GAO-05-96], 16.

74. Ibid.

75. Carole Morison, interview by author, Pocomoke City, Maryland. Ms. Morison and her husband served as contract growers for Perdue Farms.

76. Available at http://www.cia.gov/cia/publications/factbook/geos/vm.html.

77. USDA, Farm Service Agency (FSA) Fact Sheet, "Loans for Socially Disadvantaged Persons [Minorities and Women]." The other groups within this category included African Americans, American Indian, Alaskan Natives, Hispanics, and Pacific Islanders. However, affiliation to any of these groups did not automatically qualify a potential poultry grower for loan assistance. The FSA used a formula that took into account personal assets, skills, and potential to improve their skills within a given year in making a determination to grant farm loans. More information is also available at http://www.fsa.usda.gov.

78. Carole Morison interview.

79. Ibid.

80. Donald L. Barlett and James B. Steele, *America: Who Stole the Dream* (Kansas City, Missouri: Andrews and McMeel, 1996). This book initially began as a ten-part investigative report that ran in the *Philadelphia Inquirer* newspaper (November 1996). The book detailed the complex immigration laws and regulations that allowed companies including commercial poultry producers such as Perdue, Tyson, and others to import workers to work at their processing plants for minimum wages.

81. Peter Pae, "Chicken Plant Jobs Open U.S Doors for Koreans," *Washington Post*, 1 December 1999.

82. Ibid.

83. This phrase is relative. To a poor and desperate Korean immigrant, coming over to the United States to work was an opportunity of a lifetime. Any employment, regardless of how indecent by American standards that a job was judged to be, therefore constituted a decent employment for that immigrant.

84. Carole Morison interview.

85. Heather Dewar and John Rivera's article, "Poultry Plants workers find their voice," *Baltimore Sun*, 2 February 1998.

86. The Reverend Jim Lewis interview. See also, Jim Lewis, "The Church Up to Its Ears in Chicken," *The Witness Magazine* (April 2001). Available at http://thewitness.org/agw/lewis0501.html.

87. Quoted in Kevin Donahue, "Call Increases for Tyson Investigation," *Occupational Hazards: The Authority on Occupational Safety, Health and Loss Prevention*, (8 November 1999).

88. Ibid.

89. Charles N. Jeffress, "Safety and Health Professionals in the Poultry Industry." A Speech to the National Turkey Federation/National Broiler Council Safety and Health Committee (17 September 1998), 1–6.

90. See Maryland Occupational Safety and Health (MOSH), *Recordkeeping and Maryland Statistical Information,* Table A-1, Fatal Occupational Injuries by Industry and Event or Exposure, 1992–2001.

91. See "Workplace Safety and Health: Safety in the Meat and Poultry Industry . . ." 21. Critics of the commercial poultry industry such as the DPJA took advantage of any negative publicity. In this regard, commercial poultry producers were mindful of the flow of information especially when it concerned an aspect of poultry production that depicted the industry in a less than positive light.

92. According to Mr. Patrick Harmon who worked at this particular processing plant during the 1990s, Tyson Foods increased the frequency of the training of those who operated the machines in the chiller room. Incidentally, this particular plant closed permanently in 2003.

93. Robert Bussel, "Taking on 'Big Chicken'. . ." 18 and passim.

94. Bruce Drasal interview.

95. See the following press releases on http://www.ufcw.org, "Perdue Chicken Catchers Stand Up For A Voice On the Job: Catchers Demand Perdue Recognize their Right to Have a Union" [4 May 2000]; "Perdue Chicken Catchers Petition Labor Board for Union Election: Company Schemes to Evade Labor Law with Flawed Plan to Mechanize Workforce" [2 June 2000], and "Perdue Chicken Catchers Mandate: "Union Yes!" Re-Vote Nets Victory for Perdue Workers," [27 March 2001].

96. Shai Barbut, *Poultry Products Processing: An Industry Guide* (Boca Raton, Florida: CRC Press LLC, 2002), 66.

97. *UCFW Press Release*, "Perdue Chicken Catchers Petition Labor Board for Union Election: Company Schemes to Evade Labor Law with Flawed Plan to Mechanize Workforce" [2 June 2000].

98. The Annual Reports for Perdue Farms are unavailable. However, according to the available Tyson Annual Reports, the company experienced intermittent increase in sales and profits as a result of increased poultry consumption, although much of Tyson's dominant region of operation was in the South. For a more detailed examination, see Tyson Foods, Inc., Annual Reports, 1987–1997.

99. See John Bozman, "Delmarva's Competitive Edge Slipping . . ."

Chapter Seven

Conclusion

This work traced the beginnings and development of commercial poultry production on the Lower Maryland Eastern Shore up to the 1990s, and the involvement of African Americans in the industry. African Americans were mainly involved in poultry production on the labor supply side, which was crucial to the expansion of the industry. After it became commercialized in the 1930s and showed great promise in the immediate post-World War II years, poultry production expanded and became the dominant economic activity on the Lower Maryland Eastern Shore from the 1950s.

The industry expanded through innovative ways such as vertical integration, acquisitions, mergers, and consolidations. In particular, the vertical integration that began in the 1950s positioned the industry for the large-scale phase and development, which saw changes in the methods and volumes of poultry production. But the criticism of the vertical integration approach was that one company controlled all aspects of production, and therefore stifled competition with small and independent poultry producers. Subsequently, many of the small and independent poultry producers could not compete and were forced out of business.

While the expansion brought on by vertical integration was good for the poultry industry in terms of large profit margins, there were public health and environmental issues associated with the expansion. The introduction of scientific methods of breed selection and medication posed a public health concern. The bacteria-killing medications and the antibiotics fed to the chickens raised concerns about the long-term effects that these medications had on consumers. The concern was that bacteria such as *Campylobacter*, *Listeria*, *Salmonella*, *E-coli*, *Pseudomonas aeruginosa,* and others were prevalent in poultry production as it became a large-scale operation.

Studies by scientists and researchers partly centered on how to reduce the incidences of poultry-related diseases and therefore, allay public health fears. In addition, increased federal and state regulations were put in place to address the public health concerns associated with commercial poultry production. Regulations by agencies such as the Food Safety and Inspection Service (FSIS) and Food and Drug Administration (FDA) mandated poultry producers to improve food safety handling and regulated the drugs that were used in the medicated feeding of chickens. Other regulations provided for food safety inspectors to visually inspect the chickens as they came through the assembly lines. The assembly-line visual inspection was intended to reassure consumers of the government's commitment to safeguarding poultry meat from diseases and other contaminations.

Along with public health were also environmental issues. The major environmental concerns were water contamination, air pollution, and land degradation. The contention with water contamination was that the runoffs from the processing plants, which contained algae and toxins that killed fish and other aquatic life, flowed into streams and other tributaries that eventually emptied into the Chesapeake Bay. Due to commercial activities on the Bay such as fishing, crabbing, and oystering, the runoffs from the processing plants negatively affected these activities as well as other wildlife. The involvement of the Environmental Protection Agency (EPA) as a regulatory agency helped with environmental protection such as mandating companies to produce a yearly Toxics Release Inventory (TRI).

With regard to air quality, the odor that emanated from the processing plants was poor and toxic. The poor air quality was inevitable because the waste generated from processing the chickens required disposal. On the one hand, the processing plants provided job opportunities for the local community, and on the other, disposing the waste created environmental problems. This apparently contradictory situation meant that while the poultry industries were expected to maintain healthy environmental standards, the regulatory agencies had to make some concessions in order not to stifle the industries. These concessions of course drew the ire of environmental activists. Another environmental issue was land degradation. Farming remained a major part of life for many people on the Lower Maryland Eastern Shore. As waste runoffs from the processing plants made their way to the farmlands, the excessive phosphorous and nutrients sipped into the ground and had a damaging effect on the farmlands.

In spite of the public health and environmental issues, African Americans worked in the commercial poultry production as chicken catchers and processing plant workers from its inception and their labor greatly contributed to the transformation of the industry. Some of them saw working in commercial poultry as a temporary measure to provide for their economic sustenance until they were able to find better alternatives. Nevertheless, commercial poultry production was both beneficial and harsh to African Americans who worked in the industry. On the one hand, it was beneficial to African Americans in the sense that they were able to purchase their homes, automobiles, send their children to school, and provided financial support and stability for their families.

On the other hand, commercial poultry labor was harsh for African Americans. Their labor was harsh and protracted, and the working conditions in the poultry houses and processing plants were replete with health and safety hazards. They worked for low wages despite the large profit margins of the companies. In spite of their contribution to the prosperity of commercial poultry production on the Lower Maryland Eastern Shore, their role in this important sector of the economy of the Delmarva region as a whole has not been adequately documented nor has it been accorded any significant measure of recognition.

Poultry workers on the Lower Maryland Eastern Shore worked in obscurity and danger, and toiled for long hours in the chicken houses and processing plants. They were exposed to great risks of death and injuries such as repetitive motions and long-term respiratory illness. The chicken catchers in particular risked their lives as they battled chickens with sharp and extended claws, which fought off attempts to round them up before being sent to the slaughterhouses. The dangers to the catchers were further compounded as they waded through dead chickens and fecal matters to catch live chickens. In the process, they inevitably inhaled dust raised from the chickens flapping their wings (a defense mechanism), while breathing in ammonia fumes used to sanitize the chicken houses. Many of the catchers caught the chickens in enclosed houses, which further exposed them to extreme respiratory illnesses, allergies and other air-borne pathogens present in the chicken houses. These workers underwent great risks to satisfy the public's demand for poultry meat.

The dangers of commercial poultry work, as evidenced by the numerous accidents in the chicken houses and processing plants, did not deter blacks from making a living out of a most difficult occupation. They found a niche within the poultry industry by cornering the labor ranks since many of them were essentially shut out of the leadership positions in the industry. Furthermore, they did not have the financial wherewithal to own commercial poultry enterprises. The situation was not helped by the segregationist mindset of local whites, educational inequality, and inaccessibility to economic opportunities which further helped to contribute to the emergence of a black poultry labor underclass on the Lower Maryland Eastern Shore.

Particularly, in the area of education, only a small number of blacks completed high school. Had the Maryland government superimposed its power and made education more accessible to blacks, the illiteracy rate among them could have been minimized. Thus, the existence of a large poorly and relatively uneducated black labor force served the needs of large-scale commercial poultry producers. The local reality was that those, especially African Americans, who did not complete high school were more likely to be part of the poultry labor underclass, while those who completed high school and college, (many of whom were whites), were less likely to be a part of the poultry labor underclass.

From the 1930s through the 1960s, whites frustrated efforts of blacks to obtain education with measures such as educational segregation and unequal funding. Educational segregation was a weapon of choice for local whites who instituted a tiered system of labor using education as the dividing line and maintained separate schools for whites and blacks. Unequal funding was also an effective method to frustrate blacks' efforts to obtain education. Black schools were under and unequally funded, and their worth minimized, while predominantly white schools were well funded and their worth maximized.

African American labor contributed to the prosperity of the poultry companies on the Lower Maryland Eastern Shore—Mountaire Farms, Perdue Farms, and Tyson Foods. Moreover, black poultry labor also indirectly benefited the counties through the taxes paid by the poultry companies. As blacks produced and the companies sold more poultry, the county governments earned more revenues. In essence, blacks did not only work to create wealth for the companies, their labor also helped to sustain local government services.

In the final analysis, commercial poultry workers were central to the industry's financial fortunes. It is imperative for historians to recognize that the full history of commercial poultry production on the Lower Maryland Eastern Shore will be incomplete without placing African Americans in context as significant actors in both the evolutionary and transformational phases of the industry. As has often tended to be the case in many aspects of American history, African Americans have neither been duly considered as integrated participants in the wealth building of American capitalism, nor have the voices of the workers been heard. But in this instance, their contribution to Maryland and the U.S. economies are markedly evident in the sense that for over five decades, their labor greatly helped to transform the industry.

There is a dearth of secondary sources upon which to build and cross-reference some of the information obtained from oral and other primary sources. This dearth existed in this work because the poultry companies contacted for this work refused to make their records available thereby limiting access to other important primary sources. Nevertheless, significant primary source materials were obtained from the archives of advocacy and research groups, Delaware Public Archives, government departments, scientific research reports and data deposited in universities including University of Maryland, College Park, University of Maryland, Eastern Shore, Salisbury University, Johns Hopkins University, and Morgan State University, among others. The archive of the Delmarva Poultry Industry (DPI) in Georgetown, Delaware was a major source for primary and secondary sources. As a trade association that lobbies for the industry's interests, the materials from the DPI portrayed the industry in a more positive than negative light. The information from the DPI archives was therefore used with care and efforts were made to corroborate them with materials from other sources. Other important information were also obtained through oral interviews.

With the exception of Tyson Foods, which is publicly traded, the rest of the poultry companies on the Lower Maryland Eastern Shore are privately owned and are not obligated to release their records to the public or to researchers. The unavailability of important primary materials such as annual company reports limited this study in the sense that the operational details of the companies' business practices and other helpful information and data could not be fully examined. Consequently, the picture and the extent of African Americans' involvement in the privately-held commercial poultry production companies and their contributions may be incomplete. Future works in this field would extend commercial poultry historiography and could reveal more to be learned about the industry.

Another challenge to this work was the limited amount of secondary sources about African Americans on the Lower Maryland Eastern Shore in general, and the lack of scholarship that specifically showed the extent of their involvement in the commercial poultry industry in particular during its hey day. Some of the local black residents were fearful of revealing too much about their experiences with the poultry industries. They feared reprisals and possibly losing their jobs. Their fear posed a challenge in the sense that the experiences of African Americans who worked in the commercial poultry industry may not fully be known. However, the ones who shared their stories provided ample information that showed African Americans were involved in the labor supply-side of commercial poultry production on the Lower Maryland Eastern Shore. Their information was also helpful for understanding the scope and magnitude of their involvement in the industry.

Notwithstanding the above challenges, this work serves as a form of a pioneer in the comprehensive historical study of the development of commercial poultry production on the Lower Maryland Eastern Shore. It has also opened a new frontier in the labor and economic history of African Americans in the Lower Maryland Eastern Shore.

In essence, it is clear that the commercial poultry industry has been an economic mainstay of the Lower Maryland Eastern Shore with African Americans as crucial actors. This work was aimed at contributing to filling the gap that had existed about African Americans and the commercial poultry production in this important region of Maryland. Nevertheless, one implication of this pioneering work is the need for further research on commercial poultry production on the Lower Shore. Another implication is the need for studies that would focus specifically on the contributions of African Americans in other crucial economic areas on the Lower Maryland Eastern Shore. Furthermore, it is hoped that this work will help spur similar historical studies of commercial poultry production and the involvement of African Americans in the industry in other regions of the United States. It is also hoped that poultry companies would make documents such as annual reports and related data available for more studies. These annual reports are essential and would shed more light on placing the contributions of ethnic and emerging groups to the economic prosperity of regions, states, and the American society in their proper historical contexts.

Bibliography

PRIMARY SOURCES

Archives/Special Collections

Delaware Public Archives, Newark, Delaware.
National Archives of Records and Administration, College Park, Maryland.
University Archives, University of Maryland Libraries, College Park. Maryland.
The Johns Hopkins University Special Collections.
Maryland State Archives.
Nabb Research Center, Salisbury University, Salisbury, Maryland.
Morley A. Jull Collections.

Autobiography

Holden, Adele V. *Down on the Shore: The Family and Place that Forged A Poet's Voice*. Baltimore: Woodholme House Publishers, 2000.

Personal/Telephone/Oral Interviews

Beckwith, Hattie. Salisbury, Wicomico County, Maryland, 20 March 2004.

Chaloupka, George W., a retired Poultry Scientist and researcher, University of Delaware, Agricultural Experimental Service, Georgetown, Delaware, 29 November 2005.

Covell, Edward, a former poultry company owner and Executive Director of the Delmarva Poultry Industry, Inc., and National Chicken Council, Georgetown, Delaware, 29 November 2005.

Downes, Fayetta. Denton, Caroline County, Maryland, 16 March 2004.

Drasal, Bruce. Executive Vice President, United Food Commercial Workers, Local 27, telephone interview, 25 June 2004.

Grubb, Stafford Enez. Cambridge, Maryland, 15 July 2001.
Harmon, Patrick. Pocomoke City, Worcester County, Maryland, 15 March 2003.
Holden, Adele. Baltimore, Maryland, 20 January 2004.
Jackson, Spotwood. Salisbury, Maryland, telephone interview, 25 November 2004.
Lewis, Earl, The Reverend. Charleston, West Virginia, 30 July 2005.
Morton, Alison. Salisbury, Wicomico County, Maryland 19 March 2004.
Morison, Carole, a Contract grower, Perdue Farms Incorporated and Executive Director, Delmarva Poultry Justice Alliance (DPJA), Pocomoke City, Worcester County, Maryland, 12 December 2005.
"Pam," Princess Anne Local Resident, 21 March 2004.
Parvis, Connie, Director of Education and Consumer Information, Delmarva Poultry Industry, Inc., Georgetown, Delaware, 29 November 2005.
Purnell, Diana. Berlin, Worcester County, Maryland 19 March 2004.
Purnell, James L. Jr. Berlin, Maryland. Telephone interview, 25 February 2005.
Stoudmire, Eremine P. B. Salisbury, Wicomico County, Maryland, 20 March 2004.
Smith, Anna, Telephone Interview, 21 January 2004 (Baltimore-Berlin, Maryland—follow up interview, 19 March 2004).
Smith, Raymond. Berlin, Worcester County, Maryland, 17 January 2004.
Sweatt, Thomas E. Reverend, Princess Anne, Somerset County, Maryland 21 March 2004.
Truitt, Jerry, a former banker and Executive Director of the DPI, Georgetown, Delaware, 29 November 2005.

Court Cases/Documents

Abdela Tum et al. v. Barber Foods Inc., Nos. 04-66; filed in the United States Court of Appeals for the First Circuit, 1999.
Brown v. Board of Education, 347 U.S. 483 (1954), United State Supreme Court.
Dred Scott v. Sanford, 60 U.S. 393 (1856).
Gabriel Alvarez et al. v. IBP, Inc., Nos. 03-1238; filed in the United States Court of Appeals for the Ninth Circuit, 1999.
Leona Trotter et al v. Perdue Farms, Inc. [CV-99-893] Civil Action lawsuit filed at the United States District Court for the District of Delaware, 1999.
M. H. Fox et al v. Tyson Foods, Inc. [CV-99-TMP-1612-M]. Civil Action lawsuit filed at the United States District Court for the Northern District of Alabama, Middle Division, 1999.

Documentaries

CBS News. *60 Minutes*, 19 December 1999.
Delmarva Poultry Justice Alliance, *The Human Price of Poultry*. Georgetown, Delaware, 1998.
Humane Society of the United States, *Pfiesteria and the Factory Farm*. Washington, D.C.

Kerner, Ben. *Your Chicken Has Been To War*. Dover: Delaware State Archives, Division of Historical and Cultural Affairs, Hall of Records, 1943.

United States Department of Labor, E.S.A./Wage & Hour Division, *Know Your Rights*. Washington, D.C., 2001.

Commission Report

Report of the Governor's Blue Ribbon, Citizens *Pfiesteria Piscicida* Action Commission, Annapolis, Maryland, November 3, 1997.

USDA Publications/Handbooks/Bulletins

Bureau of Agricultural Economics, *Farm Production, Disposition and Income from Chickens and Eggs, 1945–1949*, [United States Department of Agriculture, BAE 268]. Washington D.C.: U.S. Government Printing Office, 1952.

Government Rules/Reports

Food and Drug Administration (HHS); "Irradiation in the production, processing, and handling of food," Final rule. *Federal Register*, 55, (May, 1990), 18538–18544.

Food Safety and Inspection Service (USDA); "Irradiation and poultry products," Proposed rule." *Federal Register*, 57, (September, 1992), 43588–43600.

Food and Drug Administration (HHS); "Irradiation in the production, processing, and handling of food," Final rule. *Federal Register*, 60, (March, 1995), 12669–12670.

Food and Drug Administration (HHS); "Irradiation in the production, processing, and handing of food," Final rule. *Federal Register*, 62, (December, 1997), 64107–64121.

"Notice of Availability of a Petition for Exemption from EPCRA and CERCLA Reporting Requirements for Ammonia from Poultry Operations," *Federal Register*, Volume 70, Number 247 (December, 2005).

United States Government Accountability Office (GAO): Report to the Minority Member, Committee on Health, Education, Labor, and Pensions, United State Senate, "Workplace Safety and Health: Safety in the Meat and Poultry Industry, while Improving, Could Be Further Strengthened," January, 2005.

United State Congress. House. Committee on Agriculture. *Agriculture Act of 1977: Report*. 95th Congress, 1st session, Report No.95-348, May, 1977.

———. *Food and Agricultural Act of 1981*: Report. 97th Congress, 1st session, *Report* No. 97-106, May 1981.

USDA *Poultry Yearbooks*, 1950–1999, and Annual Summaries.

United States Senate. A Report, "Animal Waste Pollution in America: An Emerging National Problem" compiled by the Minority Staff of the Committee on Agriculture, Nutrition, and Forestry. Senator Tom Harkin, December 1997.

Legislative Documents/Sources

Annotated Code of Maryland.
Congressional Record.
General Assembly of Maryland.
United States Senate.
United States House of Representatives.

Foundation

Shuman, Michael H. *Bay Friendly Chicken: Reinventing the Delmarva Poultry Industry.*
Maryland: Chesapeake Bay Foundation and Delmarva Poultry Justice Alliance, 2000.
Perdue, Mitzi. *Frank Perdue: Fifty Years of Building on a Solid Foundation.* Salisbury, Maryland: The Arthur W. Perdue Foundation, 1989.

Newspapers/Newsletters/Magazines Consulted

Baltimore Sun; New York Times; Washington Post; Washington Times; San Francisco Chronicle; New York Business Wire; [Salisbury] Daily Times; Black Issues of Higher Education; Las Vegas Review Journal; Rural Migration News; Agricultural Research; American Heritage; Poultry Times; Poultry & Egg Marketing; Mid-Atlantic Poultry Farmer; The Delmarva Farmer; Poultry Digest; Witness Magazine; Esquire; Feedstuffs; Fact Sheet Safety; Shore Living Magazine; Broiler Industry; Physician's Newsletter; Harper's Weekly; Public Management Magazine; Poultry Tribune, and Issues and Views.

Newspaper Opinion Commentaries and Editorials

"Contaminated Inspection Process," *The Washington Post* [Op/Ed], 25 May 1987.
"FDA Approves Antibiotic for Chickens: Drug Aimed at Preventing the Spread of Deadly E. Coli Bacteria," *The Washington Post*, 19 August 1995.
"Meat and Poultry: Inspection in Modern Times," *The Washington Post* [Op/Ed], 20 June 1987.
"Poultry Inspection Flaws Admitted," *The Washington Post*, 3 June 1987.

Press Releases

[United Food and Commercial Workers International Union]

"Baptists Back Workers In Battle With Tyson Foods: Largest African American Denomination Calls on Churches Not to Buy Tyson Products," 20 January 1999.
"Delmarva Tyson Poultry Workers Gaining Ground with New UFCW Contract," 2 February 2002.

"Food and Commercial Workers Union Calls for Quick Settlement of Charges against Tyson Foods: Union Wants Quick Action to Protect Industry from Potential Economic Disruption and Job Loss," 21 December 2001.

"Perdue Chicken Catchers Stand Up For A Voice ON the Job: Catchers Demand Perdue Recognize their Right to Have a Union," 4 May 2000.

"Perdue Chicken Catchers Petition Labor Board for Union Election: Company Schemes to Evade Labor Law with Flawed Plan to Mechanize Workforce," 2 June 2000.

"Perdue Chicken Catchers Mandate: "Union Yes!" Re-Vote Nets Victory for Perdue Workers," 27 March 2001.

Sources of Statistics/Data

Delmarva Poultry Industry, Inc.

Maryland Department of Agriculture.

Maryland Department of Planning.

Maryland Agricultural Statistical Service.

National Agricultural Statistical Service.

United States Bureau of the Census.

United States Department of Agriculture, Census of Agriculture.

Miscellaneous

"An Assessment of the Benefits of Building FibroShore: Green Energy from Poultry Litter and Forestry Residues," A Report by the *Atlantic Resource Management, Inc.*, January 2002.

Economic Situation and Prospects for Maryland Agriculture, Policy Analysis Report No. 02-01. College Park: University of Maryland, Center for Agricultural and Natural Resources Policy.

Eskin, R.A., K. H. Rowland, and D. Alegre, "Contaminants in the Chesapeake Bay Sediments 1984–1991," Chesapeake Bay Program CBP/RRS 145/96 (1996), Annapolis, Maryland.

Lichtenberg, Erick, Doug Parker, and Lori Lynch, "Economic Value of Poultry Litter Supplies In Alternative Uses," Policy Analysis Report No. 02-02, *Center for Agricultural and Natural Resource Policy*, University of Maryland, College Park, 2002.

Maryland Department of Agriculture, Maryland Agricultural Statistics, Annual Summaries, 1990–1999.

Maryland State Senate. Education, Health, and Environmental Affairs Committee. *1998 90 Day Report*, [Agriculture: Nutrient Management]. Annapolis, Maryland.

Morris, David and Jessica Nelson, "Looking Before We Leap: A Perspective on Public Subsidies for Burning Poultry Manure," *Institute for Local Self-Reliance*, October 1999.

State of the Bay, Yearly Reports, 1980–1990.

Stein, Debra, "The NIMBY Report," *National Low income Housing Coalition,* February 2001.

"Targeting Toxins: A Characterization Report," Chesapeake Bay Program. Annapolis, Maryland, 1999.

"The Broiler Industry," *A Report by the National Broiler Council*; Research 1985.

"The Poultry Industry and Water Pollution in the South," A Report, *Institute for Southern Studies*, Durham, North Carolina, December 1990.

Transcript of interview of Clarence Heath, Elwood Collins, and Lilton Sturgessi by Robert Siegel and Noah Adams on *All Things Considered,* [radio news broadcast, 8 January 1999], Washington, D.C.

Tyson Foods Inc., Annual Report, 1997.

Wakefield, Dexter B., and B. Allen Talbert, along with the references, "Exploring the Past of the New Farmers of America (NFA): The Merger with the FFA," *Proceedings of the 27th Annual National Agricultural Education Research Conference*, [n.p., n.d.], 420–433.

SECONDARY SOURCES

Books

Amenta, Edwina. *Bold Relief: Institutional Politics and the Origins of Modern American Social Policy.* Princeton: Princeton University Press, 2000.

Andrews, Matthew Page. *History of Maryland: Province and States*. Hatsboro, Pennsylvania: Tradition Press, 1965.

Baldwin, Sidney. *Poverty and Politics: The Rise and Decline of the Farm Security Administration.* Chapel Hill, North Carolina: University of North Carolina, 1968.

Barbut, Shai. *Poultry Products Processing: An Industry Guide.* Boca Raton, Florida: CRC Press LLC, 2002.

Bartlett, Donald L. and James B. Steele, *America: Who Stole the Dream.* Kansas City, Missouri: Andrews and McMeel, 1996.

Benjamin, Earl W. et al. *Marketing Poultry Products*, 5th edition. New York: John Wiley & Sons, Inc., 1960.

Biles, Roger. *The South and the New Deal.* Lexington: University Press of Kentucky, 1994.

Bonnfield, Paul. *The Dust Bowl: Men, Dirt, and Depression.* Albuquerque, New Mexico: University of New Mexico, 1978.

Bordo, Michael D., Claudia Goldin, and Eugene N. White, (Eds.), *Defining Moment: The Great Depression and the American Economy in the Twentieth Century.* Chicago: University of Chicago Press, 1998.

Botsford, Harold E. *The Economics of Poultry Management*. New York: John Wiley & Sons, Inc., 1952.

Brinkley, Alan. *The End of Reform: New Deal Liberalism in Recession and War.* New York: Vintage Books, 1996.

———. *American History: A Survey,* 10th Edition. New York: McGraw Hill College, 1999.

Brugger, Robert J. Maryland: *A Middle Temperament, 1634–1980*. Baltimore: The Johns Hopkins University Press, 1986.

Burton, Lee J. Jr., *Canneries of the Eastern Shore*. Centerville, Maryland: Tidewater Publishers, 1986.

Cantor, Milton (ed.), *Black Labor in America.* Westport, CT: Negro Universities Press, 1969.

Capper, J., G. Power, and F.R. Shivers, Jr., *Chesapeake Waters: Pollution, Public Health, and Public Opinion, 1607–1972.* Centerville, Maryland: Tidewater Publishers, 1983.

Card, Leslie E. and Malden C. Nesheim. *Poultry Production*, 11th ed. Philadelphia: Lea & Febiger, 1972.

———. *Poultry Production*, 10th ed. Philadelphia: Lea & Febiger, 1966.

Cayton, Horace R. and George S. Mitchell, *Black Workers and the New Unions.* Chapel Hill: The University of North Carolina Press, 1939.

Chapelle, Susan Ellery Green, et al. *Maryland: A History of Its People*. Baltimore: The Johns Hopkins University Press, 1986.

———, and Glenn O. Phillips. *African American Leaders: A Portrait Gallery.* Baltimore: Maryland Historical Society, 2003.

Chase, Robert A., Wesley L. Musser, and Bruce Gardner, *The Economic Contribution and Long-Term Sustainability of the Delmarva Poultry Industry*. College Park, Maryland: Center for Agricultural and Natural Resource Policy, 2003.

Cochrane, Willard W. *The Development of American Agriculture: A Historical Analysis*, 2nd ed. Minneapolis, MN: University of Minnesota Press, 1993.

"Community Economic Inventory: Wicomico County, Maryland." Annapolis, Maryland: Division of Business and Industrial Development, Maryland Department of Economic and Community Development, 1974.

Corddry, George H. *Wicomico County History*. Salisbury, Maryland: Peninsula Press, 1981.

Cronin, L.E. *Pollution in Chesapeake Bay: A Case History and Assessment. Impact of Man on the Coastal Environment.* Washington, D.C.: U.S. Environmental Protection Agency, 1982.

Davis, Karen. *Prisoned Chickens, Prisoned Eggs: An Inside Look at the Modern Poultry Plant.* Summertown, Tennessee: Book Publishing Company, 1996.

Dozier, Donald Marquand. *Portrait of the Free State: A History of Maryland.* Cambridge, Maryland: Tidewater Publishers, 1976.

Egan, Timothy. *The Worst Hard Time: The Untold Story of Those Who Survived The Great American Dust Bowl.* New York: Houghton Mifflin Company, 2006.

Ensminger, M.E. *Poultry Science*, 2nd ed. Danville, Illinois: The Interstate Printers and Publishers, Inc., 1980.

———. *Poultry Science*, 1st ed. Danville, Illinois: The Interstate Printers and Publishers, Inc. 1971.

Falk, William W. *Rooted in Place: Family and Belonging in a Southern Black Community.* Piscataway, New Jersey: Rutgers University Press, 2004.

Fannin, Mark. *Labor's Promised Land: Radical Visions of Gender, Race, and Religion in the South.* Knoxville: The University of Tennessee Press, 2003.

Faragher, John Mack, Mari Jo Buhle, Daniel Czitrom, and Susan H. Armitage, *Out of Many: A History of the American People*, Second Edition. Upper Saddle River, New Jersey: Prentice Hall Inc., 1997.

Fink, Deborah. *Cutting into the Meatpacking Line: Workers and Change in the Rural Midwest.* Chapel Hill: University of North Carolina, 1998.

Fogel, Walter A. *The Negro in the Meat Industry.* Philadelphia: University of Pennsylvania Press, 1970.

Franklin, John Hope and Alfred A. Moss Jr., *From Slavery to Slavery: A History of African Americans.* New York: McGraw-Hill Companies, Inc., 2000.

Fuke, Richard Paul. *Imperfect Equality: African Americans and the Confines of White Racial Attitudes in Post-Emancipation Maryland.* New York: Fordham University Press, 1999.

Gaines, Kevin K. *Uplifting the Race: Black Leadership, Politics, and Culture in the Twentieth Century.* Chapel Hill: The University of North Carolina Press, 1996.

Galenson, Walter. *The CIO Challenge to the AFL: A History of the American Labor Movement, 1935–1941.* Cambridge: Harvard University Press, 1960.

Gambrill, Montgomery J. *Leading Events of Maryland History with Topical Analyses, References, and Questions for Original Thought and Research.* Boston: Ginn and Company, n.d.

Gordy, Frank. *A Solid Foundation . . . The Life and Times of Arthur W. Perdue*. Salisbury, Maryland: Perdue Incorporated, 1976.

Greer, J, and D. Terlizzi, "Chemical Contamination in the Chesapeake Bay: A Synthesis of Research to Date and Future Research Directions," College Park: Maryland Sea Grant College, 1997.

Halpern, Rick and Roger Horowitz, *Meatpackers: An Oral History of Black Packinghouse Workers and their Struggle for Racial and Economic Equality.* New York: Twayne Publishers, 1996.

Harrington, Michael. *The Retail Clerks.* New York: John Willey and Sons, Inc., 1962.

Harris, Abram. *The Black Worker: The Negro and the Labor Movement.* New York: Columbia University Press, 1931.

Hine, Darlene Clark, et al. *The African American Odyssey, Since 1863,* Volume II. Upper Saddle River, New Jersey: Prentice Hall Inc., 2000.

Hoover, Herbert. *American Ideals verses the New Deal.* New York: Scribner Press, 1936.

Holden, Adele V. *Down on the Shore: A Memoir, The Family and Place that forged a Poet's Voice.* Baltimore: Woodholme House Publishers, 2000.

Horowitz, Roger. *"Negro and White, Unite and Fight!": A Social History of Industrial Unionism in Meatpacking, 1930–1990.* Urbana: University of Illinois Press, 1997.

Howard, Donald. *The WPA and Federal Relief Policy.* New York: Russell Sage Foundation, 1943.

Ifill, Sherrilyn A. *On the Courthouse Lawn: Confronting the Legacy of Lynching in the 21st Century.* Boston: Beacon Press, 2007.

Jordan, F. T. D., (ed.), *Poultry Diseases*, Third Edition. London: Bailliere Tindall, 1990.

Keatley, Kenneth J. *Place Names of the Eastern Shore of Maryland*. Queenstown, Maryland: The Queen Anne Press, 1987.

Kennedy, David M. *Freedom from Fear: The American People in Depression and War, 1929–1945*. New York: Oxford University Press, 1999.

Kerr, Allen Norwood. *The Legacy: A Centennial History of the State Agricultural Experiment Stations, 1887–1987*. Missouri: Missouri Agricultural Experiment Station, University of Missouri-Columbia, 1987.

McElvaine, Robert S. (ed.), *Down and Out in the Great Depression: Letters from the Forgotten Man*. Chapel Hill: University of North Carolina Press, 1983.

Marks, Carole C., (ed.). *A History of African Americans of Delaware and Maryland's Eastern Shore*. Wilmington, Delaware: Delaware Heritage Commission, 1998.

Marks, Robbin and Rebecca Knuffke, *America's Animal Factories: How States Fail to Prevent Pollution from Livestock Waste*. New York: Natural Resources Defense Council and the Clean Water Network, 1998.

———. *Cesspools of Shame: How Factory Lagoons and Sprayfields Threaten Environmental and Public Health*. Washington, D.C.: Natural Resources Defense Council and the Clean Water Network, 2001.

Marshall, Ray. *The Negro and Organized Labor*. New York: John Willey & Sons, Inc., 1965.

Maryland Cooperative Extension, *A Citizen's Guide to the Water Improvement Act of 1998*. Annapolis, Maryland: University of Maryland, 1998.

Maryland's Historic Somerset. Princess Anne, Maryland: Board of Education of Somerset County, 1969.

Martin, Phillip L. *Promise Unfulfilled: Unions, Immigration, and the Farm Workers*. Ithaca: Cornell University Press, 2003.

Masten, S. E., (ed.), *Case Studies in Contracting and Organization*. New York: Oxford University Press, 1996.

Moore, Joseph E., *Murder on Maryland's Eastern Shore: Race, Politics and the Case of Orphan Jones*. Charleston, South Carolina: History Press, 2006.

Newman, Wright Harry. *The Flowering of the Palatinate*. Baltimore: Genealogical Publishing Company, Inc. [reprinted] 1998.

North, Mack O. *Commercial Chicken Production Manual*. Westport, Connecticut: AVI Publishing Company, Inc., 1984.

Nourse, Edwin G., Joseph S. Davis, and John D. Black, *Three Years of the Agricultural Adjustment Administration* (Washington, D.C.: Brookings Institution, 1937.

Opie, John. *The Law of the Land: Two Hundred Years of American Farmland Policy*. Omaha: University of Nebraska Press, 1994.

Orfield, Gary and Susan E. Eaton, *Dismantling Desegregation: The Quiet Reversal of Brown v. Board of Education*. New York: The New Press, 1996.

Parkhurst, Carmen R. and George J. Mountney. *Poultry Meat and Egg Production*. New York: Van Nostrand Reinhold Company Inc., 1988.

Pozzetta, George E. *The Work Experience: Labor, Class, and Immigrant Enterprise*. New York: Garland Publishing Inc., 1991.

——. *Unions and Immigrants: Organization and Struggle.* New York: Garland Publishing Inc., 1991.

Rollo, Vera Foster. *The Black Experience in Maryland.* Lanham, Maryland: Maryland Historical Press, 1980.

——. *The Proprietorship of Maryland: A Documented Account.* Lanham: Maryland: Maryland Historical Press, 1989.

Sainsbury, David. *Poultry Health and Management*, Second edition. London: Granada Publishing Ltd. 1982.

Sawyer, Gordon. *The Agribusiness Poultry Industry: A History of Its Development.* Jericho, New York: Exposition Press Inc., 1971.

Sharpley, A., *Agricultural Phosphorous in the Chesapeake Bay Watershed: Current Status and Future Trends.* Annapolis, Maryland: Scientific and Technical Advisory Committee, Chesapeake Bay Program, 1998.

Schlesinger, Arthur Meier. *The Coming of the New Deal.* New York: Houghton Mifflin, 2003.

Shivers, George R. *Changing Times: Chronicle of Allen, Maryland, An Eastern Shore Village*. Baltimore: Gateway Press, 1998.

Skinner, John L. (ed.), *American Poultry History, 1823–1973.* Madison, Wisconsin: American Printing and Publishing, Inc., 1974.

——. *American Poultry Industry, 1974–1993*, Vol. II. Mount Morris, Illinois: Watt Publishing Company, 1996.

Smith, Page and Charles Daniel. *The Chicken Book*. Boston: Little, Brown and Company, 1975.

Steinbeck, John. *The Grapes of Wrath*, First Edition. New York: The Viking Press, 1939.

Stein, Debra. *Winning Community Support for Land Use Projects.* Washington, D.C.: Urban Land Institute, 1992.

Stern, Alice. *Poultry and Poultry-Keeping*. London: Merehurst Press, 1988.

Stromquist, Shelton and Marvin Bergman (eds.), *Unionizing the Jungles: Labor and Community in the Twentieth-Century Meatpacking Industry.* Iowa City: University of Iowa Press, 1997.

Stull, Donald D., and Michael J. Broadway. *Slaughterhouse Blues: The Meat and Poultry Industry in North America*. Belmont, California: Thomson/Wadsworth, 2004.

Stull, Donald D. et al., *Any Way You Cut It: Meat Processing and Small-Town America.* Lawrence: University Press of Kansas, 1995.

Sullivan, Patricia. *Days of Hope: Race and Democracy in the New Deal Era.* Chapel Hill: University of North Carolina Press, 1996.

Trimmer, John P. *Agricultural Maryland: A Sketch of Free State Farming.* College Park: Maryland Agricultural Extension Service, 1949.

Thompson, Charles D. and Melinda F. Wiggins, (eds.), *The Human Cost of Food.* Austin: University of Texas Press, 2002.

Thompson, Willard C. *Egg Farming: A Practical Reliable Manual Upon Producing Eggs and Poultry for Market as a Profitable Business Enterprise*. New York: Orange Judd Publishing Company, 1950.

Tobin, Bernard F., and Henry B. Arthur. *Dynamics of Adjustment in the Broiler Industry*. Boston: Division of Research, Graduate School of Business Administration, 1964.

True, A.C., and V.A. Clark, *Agricultural Experiment Stations in the United States.* Washington, D.C.: USDA, Office of Experiment Stations Bulletin 80, 1900.

United States Environmental Protection Agency, "Responses to Comments on the Notice of Proposed Rulemaking on Superfund Notification Requirements and the Adjustment to Reportable Quantities." Washington, D.C.: Environmental Protection Agency, 1985.

Wennersten, John. *Maryland's Eastern Shore: A Journey in Time and Place*. Centerville, Maryland: Tidewater Publishers, 1992.

Williams, William H. *Delmarva's Chicken Industry: 75 Years of Progress*. Georgetown, Delaware: Delmarva Poultry Industry, Inc.

Wilson, William H., and Arthur J. Burks. *The Chicken and the Egg*. New York: Coward-McCann, Inc., 1955.

Wilson, Vincent Jr., *The Book of States*. Brookeville, Maryland: American History Research Associates, 1992.

Wolters, Raymond. *Negroes and the Great Depression: The Problem of Economic Recovery.* Westport, CT: Greenwood Publisher Group Inc., 1974.

Chapters/Articles in Books

Andersen, Margaret L., "Discovering the Past/Considering the Future: Lessons from the Eastern Shore," in Carole C. Marks, (ed.), *The History of African Americans of Delaware and Maryland's Eastern Shore.* Wilmington, Delaware: Delaware Heritage Commission, 1998, 101–121.

Brock, Arnette K., "Gloria Richardson and the Cambridge Movement," in Vicki L. Crawford, Jacqueline A. Rouse, and Barbara Woods, eds., *Women in the Civil Rights Movement: Trailblazers and Torchbearers, 1941–1965* Bloomington: Indiana University Press, 1993, 121–144.

Capps, O., and J. Havlicek. "Analysis of Household Demand for Meat, Poultry, and Seafood Using the SI-Branch System," in R. Raunikar and C. L. Huang (eds.), *Food Demand Analysis: Problems, Issues, and Empirical Evidence*. Ames, Iowa: Iowa State University Press, 1987, 128–142.

Craypo, Charles. "Meatpacking: Industry Restructuring and Union Decline," in Paula B. Voos (ed.), *Contemporary Bargaining in the Private Sector*. Madison: Industrial Relations Research Association, 1994, 63–96.

Davis, Theodore J. Jr., "Socioeconomic Change: A Community in Transition," in Carole C. Marks, (ed.,), *History of African Americans of Delaware and Maryland's Eastern Shore.* Wilmington, Delaware: Delaware Heritage Commission, 1998, 199–220.

Griffith, David, "Hay Trabajo: Poultry Processing, Rural Industrialization and the Latinization of the Low-Wage Labor," in Donald D. Stull, et al., *Any Way You Cut It: Meat Processing and Small-Town America.* Lawrence: University Press of Kansas, 1995, 129–130.

Harley, Sharon, "'Chronicle of a Death Foretold': Gloria Richardson, the Cambridge Movement, and the Radical Black Activist Tradition," in Thomas, Bettye Collier-Thomas and V.P. Franklin, (eds.), *Sisters in the Struggle: African American Women in the Civil Rights–Black Power Movement.* New York: New York University Press, 2001, 174–196.

Mullikin, James C., "The Eastern Shore," in Morris L. Radoff, (ed.), *The Old Line State: a History of Maryland.* Annapolis, Maryland: Hall of Records Commission, State of Maryland, 1971, 141–169.

Small, Clara L., "Abolitionists, Free Blacks, and Runaway Slaves: Surviving Slavery on Maryland's Eastern Shore," in Carole C. Marks, (ed.), *History of African Americans of Delaware and Maryland's Eastern Shore.* Wilmington, Delaware: Delaware Heritage Commission, 1998, 55–72.

Thompson, Kathleen, "Gloria Richardson," in Darlene Clark Hines, Elsa Barkley Brown, and Rosalyn Terborg-Penn, eds., *Black Women in America: An Historical Encyclopedia,* Volume II. Brooklyn, NY: Carlson Publishing, 1993, 980–982.

Conference Presentations/Speeches

Jeffress, Charles N., "Safety and Health Professionals in the Poultry Industry." A Speech to the National Turkey Federation/National Broiler Council Safety and Health Committee, Washington, D.C., 17 September 1998.

Oscar, Thomas P. "USDA, ARS, Poultry Food Assess Risk Model (Poultry FARM), National Meeting on Poultry Health and Processing, Delmarva Poultry Industry, Inc., Ocean City, MD, 1999.

———. "Use of Computer Simulation Modeling to Predict the Microbiological Safety of Chicken," National Meeting on Poultry Health and Processing, Delmarva Poultry Industry, Inc., Ocean City, MD, 1997

Thurman, Walter N. "Have Meat Price and Income Elasticities Changed? Their Connection With Changes in Marketing Channels," in *Proceedings of the Conference on the Economics of Meat Demand.* Charleston, South Carolina, pp. 157–169., 1989.

Verg, Eric Van de (ed.,) and John H. Cumberland (ed.,), *Proceedings: Second and Third Annual Conferences on the Economics of Chesapeake Bay Management.* Annapolis, Maryland, 28–29 May [1986], and 27–29 May 1987.

Journal Articles

Aghion, P., and P. Bolton. "Contracts as a Barrier to Entry," *American Economic Review* 77 (June 1987), 388–401.

Ahmed, Ziaul Z. and Mark Sieling. "Two decades of productivity growth in poultry dressing and processing," *Monthly Labor Review Online* Volume 110, No. 4 (April 1987), 34–39.

Alaniz, Richard. "Multiple Factors Influence Declining Union Membership," *Meat & Poultry,* (May 1998), 68.

Arnesen, Eric. "Up From Exclusion: Black and White Workers, Race, and the State of Labor History," *Reviews in American History*, Volume 26, Number 1 (March, 1998), 146–174.

Aulik, Judith L. and Arthur J. Maurer. "Lactic Acid Bacteria in Poultry Products: Friend or Foe?" *Poultry and Avian Biology Review* 6 (3) 1995, 145–184.

Baker, William C., and John D. Groopman, "Introduction: Health of the Bay—Health of People Colloquium," *Environmental Research Section* A 82, (2000), 95–96.

Ball, V. E., C. Bureau, John-Christophe Bureau., R. Nehring, and A. Somwaru. "Agricultural Productivity Revisited." *American Journal of Agricultural Economics*, 79 (November, 1997), 475–486.

Ball, V. Eldon. Output, Input, and Productivity Measurement in U.S. Agriculture, 1948–1979," *American Journal of Agricultural Economics* 67 (August, 1985): 475–86.

Ball, V. E., and R. G. Chambers, "An Economic Analysis of Technology in the Meat Products Industry," *American Journal of Agricultural Economics* 62 (November, 1982), 699–709.

Banks, Taunya Lovell, "Brown at 50: Reconstructing Brown's Promise," *Washburn Law Journal*, Volume 44, No. 1 (Fall 2004), 31–64.

Barkema, Alan and Mark Drabenstott, "The Many Paths of Vertical Coordination: Structural Implications for the US Food System," *Agribusiness*, Volume 11, No. 5 (1995), 483–492.

Bird, H. R., R. J. Lillie, and J. R. Sizemore. "Environment and stimulation of chick growth by antibiotics" *Poultry Science*, 31(1952): 907.

Boehlje, Michael. "Structural Changes in the Agricultural Industries: How do we Measure, Analyze, and Understand them," *American Journal of Agricultural Economics*, Volume 81, No. 5 (November, 1999), 1028–1041.

Boesch, Donald F. "Measuring the Health of the Chesapeake Bay: Toward Integration and Prediction," *Environmental Research Section* A 82 (2000), 134–142.

Bogen, David S., "The Transformation of the Fourteenth Amendment: Reflections from the Admission of Maryland's First Black Attorneys," *Maryland Law Review*, Vol. 44 (Summer 1985).

Bowers, Douglas E. "The Research and Marketing Act of 1946 and its Effects on Agricultural Marketing Research." *Agricultural History* 56 (January 1982), 249–263.

Bouwer, H., "Returning Wastes to the Land: A New Role For Agriculture," *Journal of Soil Water Conservation*, Volume 23 (1968), 164–168.

Breuggemann, John and Cliff Brown, "The Decline of Industrial Unionism in the Meatpacking Industry . . . 1946–1987," *Work and Occupation*, (August 2003), 336.

Brown, J.C. "Broiler Financing Should be Sound," *American Poultry Journal*, 85:9 (1954).

Bugos, Glenn E. "Intellectual Property Protection in the American Chicken-Breeding Industry," *The Business History Review*, Volume 66, No. 1 (Spring, 1992), 127–168.

Bullard, W. E. Jr., "Natural Filters for Agricultural Wastes," *Soil Conservation*, Volume 34 (1968), 75–77.

Burley, Hansel, "Separate and Unequal," *American School Board Journal*, Vol. 188, No. 6 (June 2001).

Bussel, Robert. "Taking on 'Big Chicken': The Delmarva Poultry Justice Alliance," *Labor Studies Journal*, Volume 28, No. 2 (Summer 2003), 1–24.

Callcott, George H. *History of the University of Maryland.* Baltimore: Maryland Historical Society, 1966.

Cartenson, Vernon. "the Genesis of an Agricultural Experiment Station." *Agricultural History* 34 (January 1960), 13–20.

Cassedy, Gilbert James. "African Americans and the American Labor Movement," *Prologue: Quarterly of the National Archives and Records Administration* Volume 29, No. 2 (Summer 1997), 113–121.

Chapman, H.D., Z.B. Johnson, and J.L. McFarland, *Poultry Science*, Volume 82 (2003), 50–3.

Coates, M. E., C. D. Dickerson, G. F. Harrison, S. K. Kon, S. H. Cummins, and W. F. J. Cuthberton. "Mode of antibiotics in stimulating growth in chicks" *Nature*, 168 (1951), 332.

Corey Reece R. and Joseph M. Byrnes. "Oxytetracycline-Resistant Coliforms in Commercial Poultry Products" *Applied Microbiology*, Volume 11, No. 6 (November, 1963), 481–484.

Dalfiume, Richard M. "The Fahy Committee and Desegregation of the Armed Forces." *Historian* 31 (November 1968), 1–20.

Datta, S., H. Niwa, and K. Itoh, "Prevalence of 11 pathogenic genes of Campylobacter *jejuni* by PCR in Strains isolated from humans, poultry meat and broiler and bovine faeces," *Journal of Medical Microbiology*, Volume 52, No. 4 (April 2003), 345–348.

Davidson, Thomas E., "Free Blacks in Old Somerset County, 1745–1755," *Maryland Historical Magazine*, Volume 80, No. 2 (Summer 1985), 151–156.

deFur, Peter, and Lisa Foersom, "Toxic Chemicals: Can What We Don't Know Harm Us?" *Environmental Health Section* A 82, (2000), 113–133.

Donahue, Kevin. "Call Increases for Tyson Investigation," *Occupational Hazards: The Authority on Occupational Safety, Health and Loss Prevention*, (8 November 1999).

Dutka, B.J., and J.B. Bell, "Isolation of *Salmonella* from Moderately Polluted Waters," *Journal of Water Pollution Control Federation*, Volume 45, No. 2 (February 1973), 316–324.

Dutkiewicz, Jacek, "Exposure to Dust-Borne Bacteria in Agriculture. I. Environmental Studies," *Archives of Environmental Health*, Volume 33, Issue 5 (September/October 1978), 250–259.

Fehn, Bruce. "African American women and the struggle for equality in the meatpacking industry, 1940–1960," *Journal of Women's History*, 10:1 (1998), 45–69.

Frye, G. R., H. H. Weise, and A. R. Winter. "Relative effectiveness of increasing shelf life of poultry meat by long and short period of antibiotic feeding" *Food Technology*, 12 (supplement):52 (1958).

Gerald, J. O., "Influence of Trade Credit," *Journal of Farm Economics*, 37, 5 (1955).

Gordon, R. F., J. S. Garside, and J. F. Tucker. "Emergence of resistant strains of bacteria following the continuous feeding of antibiotics to poultry" *International Veterinary Congress, Madrid*, 2 (1959), 347–349.

Harvey, S., R.A. Fraser, and R.W. Lea, "Growth hormone secretion in Poultry," *Critical Reviews in Poultry Biology*, Volume 3, No. 4 (1991), 239–282.

Hoadley, A. W., W.M. Kemp, A.C. Firmin, G.T. Smith, and P. Schelhorn, "*Salmonellae* in the Environment Around a Chicken Processing Plant," *Applied Microbiology*, Volume 27, No. 5 (May 1974), 848–857.

Hooker, Neal et al. "Communicating Food Safety Gains," *American Journal of Agricultural Economics* Volume 81, No. 5 (1999 Proceedings) 1102-1106. 1997.

Horrigan, Leo, Robert S. Lawrence and Polly Walker, "How Sustainable Agriculture can Address the Environmental and Human Health Harms of Industrial Agriculture," *Environmental Health Perspectives*, Volume 110, Number 5 (May, 2002), 445–456.

Horowitz, Roger, "Be Loyal to Your Industry": J. Frank Gordy, Jr., the Cooperative Extension Service, and the Making of a Business Community in the Delmarva Poultry Industry, 1945–1970," *Delaware History* Volume XXVII, Nos. 1–2 (Spring-Summer 1996 Fall-Winter 1996–7), 3–18.

———, and Rick Halpern, "Work, race, and identity: self-representation in the narratives of black packinghouse workers" *Oral History Review*, 26:1(1999), 23–43.

Hoyt, William D., Jr., "Two Maryland Farm Leases," *Agricultural History* Volume 21 (1947), 185–187.

Innes, Robert, and Richard J. Sexton, "Strategic Buyers and Exclusionary Contracts, *American Economic Review* 84 (June, 1994), 566–584.

Ippolito, Pauline M., and Alan D. Mathios, "The Regulation of Science-Based Claims in Advertising," *Journal of Consumer Policy* 13:413–445.

Ismond, Alan, and P. Eng, "Irradiation's not the Solution," *Letters, Meat & Poultry*, (December 1995), 4.

Janke, Katelan, *Survival in the Storm: The Dust Bowl Diary of Grace Edwards, Dalhart, Texas, 1935, Scholastic* (September 2002).

Jeffry, F.P., "Basis for Credit," *American Poultry Journal*, 85:1:14.

Jones, Lu Ann and Nancy Grey Osterud, "Breaking New Ground: Oral History and Agricultural History," *Journal of American History* Volume 76 (September 1989), 551–564.

Kimmel, Ross M., "Free Blacks in Seventeenth Century Maryland," *Maryland Historical Magazine*, Volume 71, No. 1 (Spring 1976), 19–25.

Klein, B., R. Crawford, and A. Alchian, "Vertical Integration, Appropriate Rents and Competitive Contracting Process," *Journal of Law and Economics*, Vol. 21, No. 2, pp. 297–326.

Kotlowski, Dean. "The Last Lonely Shore: Nature, Man, and the Making of Assateague Island National Seashore," *Maryland Historical Magazine* Volume 99, No. 2 (Summer 2004), 165–195.

Kazwala, R.R, Collins J.D., Hannan J. Crinion R.A, O'Mahony H. "Factors responsible for the spread of campylobacter jejuni infection in commercial poultry," *Veterinary Record*, 126(13) (March 31 1990), 305–6.

Kimberly, Charles M, "The Depression in Maryland: The Failure of Voluntaryism," *Maryland Historical Magazine*, Vol. 70, No. 2 (Summer 1975), 189–202.

Knoeber, Charles R. "A Real Game of Chicken: Contracts, Tournaments, and the Production of Broilers," *Journal of Law, Economics, and Organization* 5 (2), Fall 1989.

———, et al. "Don't Count Your Chickens . . .': Risk and Risk Shifting in the Broiler Industry," *American Journal of Agricultural Economics* 77 (August, (1995), 486–496.

Kraft, D. J., et al. "Salmonella in Wastes Produced at Commercial Poultry Farms" *Applied Microbiology*, Volume 18, No. 5 (November, 1969), 703–704.

Kuebler, Edward J., "The Desegregation of the University of Maryland," *Maryland Historical Magazine*, Volume 71, No. 1 (Spring 1976), 37–49.

Lilburn, M. S. "Skeletal growth of commercial poultry species," *Poultry Science*, 73(6):897–903. (1994).

Maher, Timothy, "Environmental Oppression: Who Is Targeted for Toxic Exposure?" *Journal of Black Studies,* Volume 28, No. 3 (January, 1998), 357–367.

Mameesh, M. S., B. Sass, and B. C. Johnson, "The assessment of the antibiotic growth response in the chick" *Poultry Science*, 38 (1959), 512–515.

Max, Donald, D., "Black Immigrants: The Slave Trade in Colonial Maryland," *Maryland Historical Magazine*, Volume 73, No. 1 (March 1978), 31–45.

Millner, Sandra, "Recasting Civil Rights Leadership: Gloria Richardson and the Cambridge Movement," *Journal of Black Studies*, 26 (July 1996), 668–687.

Morris, George K., Joy G Wells, *Salmonella* Contamination in a Poultry-Processing Plant," *Applied Microbiology*, Volume 19, No. 5 (May 1970), 795–799.

Morris, P. et al., "Respiratory Symptoms and Pulmonary Function in Chicken Catchers in Poultry Confinement Units," *American Journal of Industrial Medicine*, Volume 19 (1991), 195–204.

Morris, J. Glen Jr., "Learning and Memory Difficulties after Environmental Exposure to Waterways Containing Toxin-Producing *Pfiesteria* or *Pfiesteria*-like Dinoflagellates," *The Lancet*, Volume 352, Issue 9127 (15 August 1998), 523–539.

Mormino, Gary R., "GI Joe Meets Jim Crow: Racial Violence and Reform in World War II Florida," *Florida Historical Quarterly* 73 (July 1994), 23–42.

Nelson, Phillip, "Information and Consumer Behavior," *Journal of Political Economy* 78(2), 311–329.

Newell, D. G., and C. Fearnley, "Sources of Campylobacter Colonization in Broiler Chickens," *Applied and Environmental Microbiology*, Volume 69, No. 8 (August 2003), 4343–4351.

Neville, Barry and Edward Jones, "Slavery in Worcester County, Maryland, 1688–1766," *Maryland Historical Magazine*, Volume 89, No. 3 (Fall 1994), 319–327.

Omo-Osagie, Solomon Iyobosa II, "'Count Her In' ": Enez Stafford Grubb in the Building and Rebuilding of Cambridge, Maryland," *Southern Historian: A Journal of Southern History* Volume XIV (Spring 2003), 40–49.

Oscar, P. Thomas, "The Development of a risk assessment model for use in the poultry industry," *Journal of Food Safety* 18 (1998) 371–381.

———."Development and Validation of Primary, Secondary, and Tertiary Models for Growth of Salmonella Typhimurium in Sterile Chicken," *Journal of Food Protection*, Volume 68, No. 12 (2005), 2606–2613.

———. "Validation of Lag Time and Growth Rate Models for Salmonella Typhimurium: Acceptable Prediction Zone Method," *Journal of Food Science*, Volume 70, No. 2 (2005), 129–137.

———. "Simulation Model for Enumeration of Salmonella of PCR Detection Time Score and Sample Size: Implications for Risk Management," *Journal of Food Protection*, Volume 67, No. 6. (2004), 1201–1208.

Parker, Doug, "Alternative Uses of Poultry Litter," *Economic Viewpoints*, Volume 3, No. 1 (Summer, 1998).

Philips, Glenn O., "Maryland and the Caribbean, 1634–1984: Some Highlights," *Maryland Historical Magazine*, Vol. 83, No. 3 (Fall, 1988), 199–214.

Rasmussen, Eric B., J. Mark Ramseyer, and John S. Wiley, Jr. "Naked Exclusion," *American Economic Review* 81 (December, 1991), 1137–1145.

Reading, D.C., "New Deal Activity and the States," *Journal of Economic History*, Volume 36 (December, 1973), 792–810.

Ricklefs, R. E., "Modification of growth and development of muscles of poultry," *Poultry Science*, Volume 64 (1985), 1563–1576.

Rooson, C. P., M. D. Hammig, and J. W. Jones. "Foreign Market Promotion Programs: An Analysis for Apples, Poultry, and Tobacco." *Agribusiness: An International Journal* 2 (1986), 33–42.

Roy, Ewell Paul. "Effective Competition and Changing Patterns in Marketing Broiler Chickens," *Journal of Farm Economics* Volume 48, Issue 3 (August 1966), 188–201.

Schubert, Siegfried D., Max J. Suarez, Philip J. Pegion, Randal D. Koster, and Julio T. Bacmeister, "On the Cause of the 1930s Dust Bowl," *SCIENCE* Magazine, (19 March 2004), 1855–1859.

Schultz, Constance B. "Maryland Fifty Years Ago: Remembering the CWA and WPA," *Maryland Heritage News*, 2 (Fall 1984), 8–9.

Scanes, Collin G. "Chickens—A Model for Growth?" *Critical Review Poultry Biology* 3(1991), 225–227.

Sebold, Kimberly R. "The Delmarva Broiler Industry and World War II: A Case in Wartime Economy," *Delaware History* Volume XXV, No. 3 (Spring-Summer, 1993), 200–214.

———, "Chicken-House Apartments on the Delmarva Peninsula," *Delaware History* Volume XXV, No. 4 (Fall-Winter, 1993–4), 253–263.

Sexton, Brendan. "The Working Class Experience," *The American Economic Review*, Volume 62, No. 1/2 (1972), 149–153.

Shane, Simon M. "Factors Influencing Health and Performance on Poultry on Hot Climates." *CRC Critical Reviews in Poultry Biology*, Volume 1, Issue 3 (1988), 263.

———. "Campylobacter infection of commercial poultry," *Revue Scientifique et Technique*, 19(2): 2000, 376–95.

Shusterman, Dennis, "Critical Review: The Health Significance of Environmental Odor Pollution," *Archives of Environmental Health*, Volume 47, No. 1 (January/February, 1992), 76–87.

Simpson, T. W. "Agronomic Use of Poultry Industry Waste," *Poultry Science* 70 (1991), 1126–1131.

Smil, Vaclav, "Phosphorous in the Environment: Natural Flows and Human Interferences," *Annual Reviews of Energy and the Environment*, 25 (2000), 53–88.

Smith, Richard Norton and Timothy Walch, "The Ordeal of Herbert Hoover," *Prologue: Quarterly of the National Archives and Records Administration* Volume 36, No. 2 (Summer, 2004), 30–38.

Smith, Sandy, "Perdue Farm Settles, Tyson Foods Fights Donning and Doffing Disputes," *Occupational Hazards: The Authority on Occupational Safety, Health and Loss Prevention*, (10 May 2002).

Sorisio, Paul L. "Poultry, Waste, and Pollution: The Lack of Enforcement of Maryland's Water Quality Improvement Act." *Maryland Law Review* 62 (no. 4, 2003), 1054–1075.

Spradbrow, P.B. "Newcastle Disease in Village Chickens." *Poultry Science Review* 5 (1993/4), 57–96.

Spino, Donald F. "Elevated-Temperature Technique for the Isolation of Salmonella from Streams" *Applied Microbiology*, Volume 14, No. 4 (July, 1966), 591–596.

Stein, Debra, "The Ethics of NIMBYism," *Journal of Housing and Community Development* (November/December, 1996), Available at http://www.nahro.org/publications/johcd.cfm.

Striffler, Steve. "Inside a Poultry Processing Plant: An Ethnographic Portrait," *Labor History*, Volume 43, No. 3 (2002), 305–313.

U.S. Department of Commerce, Census Bureau. *Poultry Processing*, 1997. 1997 Economic Census of Manufacturing Industry Series EC97M-3116A. 1999b.

Ventsias, Tom, "Waste Not, Want Not: Putting Delmarva's Poultry Litter to Good Use," *Maryland Research*, Volume 11, Number 2 (Spring, 2002). Available at http://www.marylandresearch.umd.edu.

Vukina, T., and W. E. Foster, "Efficiency Gains in Broiler Production Through Contract Parameter Fine Tuning," *Poultry Science*, Volume 75 (November, 1996), 1351–1358.

Walker, Polly and Robert S. Lawrence, "American Meat: A Threat to your Health and to the Environment" *Yale Journal of Health Policy, Law, and Ethics*, Volume IV, Issue 1 (Winter, 2004), 173–182 [book review].

Walker, Homer W and John C. Ayres, "Incidence and Kinds of Microorganisms Associated with Commercially Dressed Poultry," *Applied Microbiology*, Volume 4, No. 6 (November, 1956), 345–349.

Wallis, John Joseph, "Employment, Politics, and Economic Recovery During the Great Depression," *The Review of Economics and Statistics*, Volume 59 (August, 1987), 516–520.

Wakabayashi, Muneo, Betsy G. Bang, and Frederick B. Bang, "Mucociliary Transport in Chickens with Newcastle Disease Virus and Exposed to Sulfur Dioxide," *Archives of Environmental Health*, Volume 32, Issue 3 (May/June, 1977), 101–108.

Westgren, R. E. "Case Studies of Market Coordination in the Poultry Industries." *Canadian Journal of Agricultural Economics*, Vol. 42, No. 4, December, 1994), 565–575.

Wennersten, John R., "A Cycle of Race Relations on Maryland's Eastern Shore: Somerset County, 1850–1917," *Maryland Historical Magazine*, Volume 80, No. 4 (Winter, 1985), 377–382.

———, and Ruth Ellen Wennersten, "Separate and Unequal: The Evolution of a Black Land Grant College in Maryland, 1890–1930," *Maryland Historical Magazine*, Volume 72, No. 1 (Spring, 1977), 110–117.

Wilder, ANN, R.A. MacCready, "Isolation of *salmonella* from poultry, poultry products and poultry processing plants in Massachusetts," *New England Journal of Medicine*, Volume 274, No. 26 (30 June 1966), 1453–1460.

Wiser, Vivian, "Meat and Poultry Inspection in the United States Department of Agriculture," Vivian Wiser, Larry Mark, and H. Graham Purchase (eds.). *Journal of NAL Associates* (11): (January-December, 1986), 4–5.

Yacowitz, H. "Antibiotic levels in the digestive track of the chick," *Poultry Science*, 32 (1953), 966–968.

Yoon, K.S., and T. P. Oscar, "Survival of Salmonella Typhimurium of Sterile Ground Chicken Breast Patties After Washing with Salt and Phosphates and During Refrigerated and Frozen Storage," *Journal of Food Science*, Volume 67, No. 2 (2002), 772–775.

———, and C. N. Burnette, and T. P. Oscar, "Development of Predictive Models for the Survival of Campylobacter *jejuni* (ATCC 43051) on Cooked Chicken breast Patties and in Broth as a Function of Temperature," *Journal of Food Protection*, Volume 67, No. 1 (2004), 64–70.

Young, James T., "Origins of New Deal Agricultural Policy: Interest Group's Role in Policy Formation," *Policy Studies Journal*, Volume 21, Issue 2 (June, 1993), 190–209.

USDA Publications/Handbooks/Bulletins

Ackerman, K. Z., M. E. Smith, and N. R. L. Suarez. "Agricultural Export Programs, Background for the 1995 Farm Legislation." *Agricultural Economics Report No. 716*. Washington, D.C.: USDA/Economic Research Service, 1985.

Benson, Verel W., and Thomas J. Witzig., "The Chicken Broiler Industry: Structure, Practices, and Costs," *Agricultural Economic Report No. 381*. Washington, D.C.: United States Department of Agriculture, Economic Research Service, 1977.

Bunch, Karen, [compiler]. *Food, Consumption, Prices, and Expenditures, 1963–83*. Washington, D.C.: USD/Economic Research Service, [USDA Statistical Bulletin no. 713], 1984.

Coyle, Barry, Robert G. Chambers, and Andrew Schmitz. *Economic Gains from Agricultural Trade: A Review and Bibliography*. Washington, D.C.: USDA/ Economic Research Service, [USDA Bibliographies and Literature on Agriculture no 48], 1986.

Crom, Richard J. *Economics of the U.S. Meat Industry*. U.S. Department of Agriculture. Economic Research Service. AIB-545. November, 1988.

Crutchfield, Steven R et al. *Economic Assessment of Food Safety Regulations: The New Approach to Meat and Poultry Inspection* AER-755. USDA/ Agricultural Economic Research Service, 1997. Available at www.ers.usda.gov/publications.aer755.

Daberkow, Stan G., and Leslie A. Whitener. *Agricultural Labor Date Sources: An Update*. Washington, D.C.:USDA/Economic Research Service, [USDA Agricultural Handbook no. 658], 1986.

Glaser, Lawrence K. *Provisions of the Food Security Act of 1985*. Washington: USDA, Economic Research Service Agriculture [Information Bulletin Number 498], 1986.

Golan, Elise et al. *Economics of Food Labeling*. USDA Agricultural Economic Report No. 793, December, 2000.

Hamilton, A.B., and C.K. McGee, "The Economic and Social Status of Rural Negro Families in Maryland." (College Park, Maryland: The University of Maryland Agricultural Experiment Station and Extension Service, 1948). [*Bulletin X4*]

Henson, William L. "The U.S. Broiler Industry: Past, Present Status, Practices, and Cost," *Agricultural Experiment and Research Services*, USDA, No. 149 (May, 1980).

Ingalshe, Gene. *Cooperative Facts*. Washington, D.C.: U.S. Department of Agriculture, Economics, Statistics, and Cooperatives Services, [Agricultural Cooperative Service Information Report no. 2, (H182)], 1987.

Lasley, Floyd A., William L. Henson, and Harold B. Jones. *The U.S. Poultry Industry: Changing Economics and Structure*. USDA Agricultural Economic Report No. 502, February, 1983.

MacDonald James M. et al. *Consolidation in U.S. Meatpacking*. USDA Agricultural Economic Report No. 785, February, 2000.

———. *Contracts, Markets, and Prices: Organizing the Production and Use of Agricultural Commodities*. USDA Agricultural Economic Report Number 837, November, 2004.

Marrion, B.W., and H. B. Arthur. *Dynamic Factors in Vertical Commodity Systems: A Case Study of the Broiler System.* Research Bulletin No. 1065, Ohio Agricultural Research and Development Center, Ohio State University, 1973.

Martinez, Steve W. *Vertical Coordination in the Pork and Broiler Industries: Implications for Pork and Chicken Products*. USDA Agricultural Economic Report No. 777, April, 1999.

Ollinger, Michael et al. *Meat and Poultry Plants' Food Safety Investments: Survey Findings.* USDA Technical Bulletin Number 1911, May, 2004.

Ollinger, Michael, and Valerie Mueller. *Managing for Safer Food: The Economics of Sanitation and Process Controls in Meat and Poultry Plants*. AER-817. USDA, Economic Research Service, 2003. Available at www.ers.usda.gov/publications/aer817.

Reddington, John. "Diversity is Key to continued success of U.S. Poultry Exports" USDA/FAS, Washington, DC. Available at: http://www.fas.usda.gov:80/info/agexporter/1997/poultry.html.

Reidmund, Donn A., J.R. Martin and C. V. Moore. *Structural Changes in Agriculture: The Experience for Broilers, Fed Cattle, and Processing Vegetables*. USDA Technical Bulletin No. 1648, 1980.

Scientific/Technical Articles

Carter, T. A. and L. E. Carr, Poultry Hatchery Effluent Quality," Scientific Article No. A2111, Contribution No. 5068, Maryland Agricultural Experiment Station, Department of Poultry Science, University of Maryland, College Park.

Gardiner, Walter H., and John W. Wysong, "Combining Farm and NonFarm Labor Employment Opportunities," *Agricultural and Resource Economics Information Series*, [No. A-1], 1–2; Box 8, Contribution Nos., 5000–5249, Maryland Agricultural Experiment Station, October, 1975.

Lynk, William A., F. Howard Forsyth, and Virginia S. Krohnfeldt, "Characteristics of Families in Poverty in a Rural Maryland County," [Scientific Article A1679]. College Park: University of Maryland, College Park; Princess Anne: University of Maryland, Eastern Shore, and Agriculture Experiment Station, June, 1971.

Miller, Gary L., and John H. Axley, "Poultry Processing Plant Wastewater Disposal on SOD," Technical Report No. 36, [Completed Report A-013-Md 14-31-0001-5020], *Water Resources Research Center*, University of Maryland, College Park (July, 1969–1975).

Ngo, Axuan, and G. R. Beecher, "Dietary Glycine Requirements for Growth and Cellular Development in Chicks," Scientific Article No. A2247, Contribution No. 5237, Maryland Agricultural Experiment Station, Department of Poultry Science, University of Maryland, College Park.

Twining, P. V., Jr., O. P. Thomas, and E. H. Bossard, "The Number of Feathers on the Litter: Another Criterion for Evaluating the Adequacy of Broiler Diets," Scientific Article No. A2139, Contribution No. 5103, Maryland Agricultural Experiment Station, Department of Poultry Science, University of Maryland, College Park.

Theses/Dissertations

Beiter, Joseph Robert. "An Analysis of Farmer Cooperatives in Maryland." M.S. thesis, University of Maryland, College Park, 1957.

Coddington, James W. "Farm Management Study of 266 Farms on the Eastern Shore of Maryland. 1931." M.S. thesis, University of Maryland, College Park, 1933.

Davies, Thomas J. "The Broiler Industry in Maryland." M.S. thesis, University of Maryland, College Park, 1942.

Dyne, Van Donald L., "Economic Feasibility of Heating Maryland Broiler Houses with Solar Energy." Ph.D. diss., University of Maryland, College Park, 1976.

Gangle, Brian James, "Sources and Occurrence of Antibiotic Resistance in the Environment." M.S. thesis, University of Maryland, College Park, 2005.

Goodwin, Frank. "A study of Personal and Social Organization: An Explorative Survey of the Eastern Shore of Maryland." Ph.D. diss., University of Pennsylvania, 1944.

Gunter, Lewell F. Jr., "The Optimal Placement of Breeder Flocks in an Integrated Broiler Firm." Ph.D. diss., University of Maryland, College Park, 1979.

Halpern, Eric Brian, "'Black and White Unite and Fight': Race and Labor in Meatpacking, 1904–1948." Ph.D. diss., University of Pennsylvania, 1989.

Hayes, Joshua Richard, "Multiple Antibiotic Resistances of *Enterococci* from the Poultry Production Environment and Characterization of the Macrolide-Lincosamide-Streptogramin Resistance Phenotypes of Enterococcus Faecium." Ph.D. diss., University of Maryland, College Park, 2004.

Hauver, W. E., Jr. "An Economic Study of 99 Farms in Maryland." M.S. thesis, University of Maryland, College Park, 1935.

Horowitz, Roger. "The Path Not Taken: A Social History of Industrial Unionism in Meatpacking, 1930–1960." Ph.D. diss., University of Wisconsin-Madison, 1990.

Jackson, Cezar Tampoya. "A Comparative Study of Perceptions of the Media Relating to Lynchings on the Eastern Shore of Maryland." M.A. thesis, Salisbury State University, 1996.

Kimberly, Charles M. "The Depression and New Deal in Maryland." Ph.D. diss., American University, 1974.

Lafourcade, Olivier. "Analysis of Optimal Marketing Strategies for the Poultry Industry in Delmarva." Ph.D. diss., University of Maryland, College Park, 1971.

Larsen, V. "Agricultural Wastewater Rectification from a Poultry Processing Plant." M.S. thesis, University of Maryland, College Park, 1970.

Levin, James B. "Albert C. Ritchie: A Political Biography." Ph.D. diss., City University of New York, 1970.

Lodman, William J., "Farm Credit on the Lower Eastern Shore of Maryland." M.S., thesis, University of Maryland, College Park, 1941.

McCarron, Edward Gerard. "Governor Albert C. Ritchie and Unemployment Relief in Maryland, 1929–1933." M.A. thesis, University of Maryland, College Park, 1969.

Paul R. Poffenberger, "An Economic Study of the Broiler Industry in Maryland." M.S. thesis, University of Maryland, College Park, 1937.

Pierce, C. W., "An Economic Study of 99 Maryland Poultry Farms." M.S. thesis, University of Maryland, College Park, 1933.

Poffenberger, Paul R., "An Economic Study of the Broiler Industry in Maryland." M.S., thesis, University of Maryland, 1937.

Reid, George Edward Jr., "An Analysis of the Market Organization and Structure of the Maryland Table Egg Industry." M.S thesis, University of Maryland, College Park, 1966.

Rogers, George Burnet, "Economies of Scale in Plants Producing Further-Processed Poultry Products and Resulting Impacts." Ph.D. diss., University of Maryland, College Park, 1966.

Stiles, John Stanley Jr., "Comparative Costs of Cutting and Packaging Poultry." M.S. thesis, University of Maryland, College Park, 1959.

Trever, Edward K. "Gloria Richardson and the Cambridge Civil Rights Movement, 1962–1964." M.A. thesis, Morgan State University, 1994.

Wheeler, Radcliffe W. "Rural Cooperative Credit in Maryland." M.S. thesis, University of Maryland, College Park, 1957.

Yonkos, Lance Thompson. "Poultry Litter-Induced Endocrine Disruption: Laboratory and Field Investigations." Ph.D. diss., University of Maryland, College Park, 2005.

Unpublished and Undated Papers, Pamphlets, and Brochures

Chossek, Kristen, Polly Walker, Thomas A. Burke, and Beth Resnick, "The Chesapeake Bay Health Indicators Project: Linking Ecological and Human Health," (The Center for a Livable Future, Johns Hopkins Bloomberg School of Public Health, Baltimore, Maryland) [unpublished paper and n.d.].

Chaloupka, George W, "The Early Days of our Poultry," Georgetown, Delaware (n.d.).

Covell, Edward H., "The Broiler Industry—Then, Now and Tomorrow," Unpublished paper.

Eisenberg, Deborah Thompson, "The Feudal Lord in the Kingdom of Chicken: Contracting and Worker Exploitation by the Poultry Industry," *The Public Justice Center*, Unpublished paper and n. d.

"How it all Began," *Introduction to Perdue Farms* [Perdue brochure].

Rogers, Richard T., "Broilers—Differentiating a Commodity," Unpublished paper.

"Science Overview: Public Health Implications of Industrial Animal Production," *Johns Hopkins Bloomberg School of Public Health, Baltimore, Maryland.* [pamphlet, n.d.].

Stein, Holly A., "Evolution of the Broiler Industry on the Eastern Shore," 17 March 1986. Unpublished paper.

"The Delmarva Peninsula—Birthplace of Commercial Broiler Industry," Delmarva Poultry Industry, Inc. [pamphlet—n. d.]

Wallis, John, Price Fishback, and Shawn Cantor, "Politics, Relief, and Reform: The Transformation of America's Social Welfare System during the New Deal," Unpublished paper and n.d.

Web Sites

American Federation of Labor-Council of Industrial Organizations http://www.afl-cio.org.

Bureau of Labor Statistics (BLS) http://www.bls.gov.

Centers for Disease and Control http://www.cdc.gov.

Chesapeake Bay Foundation http://www.cbf.org.

Delaware Online http://www.delawareonline.com.

Delmarva Poultry Industry http://www.dpichicken.org.

Food and Drug Administration http://www.fda.gov/cvm.

Human Rights Watch http://www.hrw.org.

Interfaith Worker Justice http://www.nicwj.org.
Keep Antibiotic Working http://www.keepantibioticworking.com.
Labor Research Association http://www.laborresearch.org.
The Lower Shore http://www.lowershore.net/chickens.
Maryland Occupational Safety and Health (MOSH) http:///www.mosh.state.md.us.
Maryland State Archives http://www.mdarchives.state.md.us/agriculture/poultry.
National Association of Housing and Redevelopment Officials http://www.nahro.org/publications/johcd.cfm.
Occupational Safety and Health Administration (OSHA) http://www.osha.gov.
Farm Sanctuary http://www.poultry.org/labor.
Retail, Wholesale and Department Store Union http://www.rwdsu.org.
Sierra Club http://www.sierraclub.org/factoryfarms/report99/perdue.asp.
Union of Concerned Scientists http://www.ucsusa.org.
United Food and Commercial Workers International Union http://www.ufcw.org.
United Poultry Concerns http://www.upc-online.org.
United States Department of Agriculture http://www.usda.gov.
United State Department of Labor http://www.dol.gov.

CPSIA information can be obtained at www.ICGtesting.com
Printed in the USA
BVOW070756300512

291167BV00001B/3/P